Theories of High Temperature Superconductivity

J. Woods Halley, Editor

School of Physics and Astronomy
University of Minnesota, Minneapolis

Addison-Wesley Publishing Company, Inc.
The Advanced Book Program
Redwood City, California • Menlo Park, California
Reading, Massachusetts • New York • Amsterdam
Don Mills, Ontario • Sydney • Bonn • Madrid • Singapore
Tokyo • San Juan • Wokingham, United Kingdom

PHYSICS

Publisher: *Allan M. Wylde*
Production Administrator: *Karen L. Garrison*
Editorial Coordinator: *Pearline Randall*
Electronic Production Consultant: *Mona Zeftel*
Promotions Manager: *Celina Gonzales*

This volume was typeset using AMS-T$_E$X on an Apollo computer. Camera-ready output from an Apple Laser Writer Plus Printer.

Copyright © 1988 by Addison-Wesley Publishing Company

All rights reserved. No part of this publication may reproduced, stored in a retrieval system, or transmitted in any form or by any means, electronic, mechanical, photocopying, recording, or otherwise, without the prior written permission of the publisher. Printed in the United States of America. Published simultaneously in Canada.

ISBN 0-201-12008-9
ABCDEFGHIJ-AL-898

Contents

QC 612
S8 T481
1988
PHYS

BACKGROUND

1   Introduction
      J. W. Halley   3

2   Historical Introduction
      John Bardeen   7

3   Review of Experiments
      Allen Goldman   13

BCS-LIKE THEORIES

4   Theories of High T_c Superconductors
      John Bardeen   29

5 Role of Interlayer Coupling in Oxide Superconductors

Zlatko Tesanovic   35

6 Electronic Structure and Properties of Quasi-Two-Dimensional Layered Superconducting Perovskites: $La_{2-x}M_xCuO_4(M = Ba,Sr...)$

A. Freeman, Jaejun Yu and C. L. Fu   43

7 Electronic Structure, Charge Transfer Excitations and High Temperature Superconductivity

A. J. Freeman, Jaejun Yu, S. Massidda and D. Koelling   53

8 Theoretical Calculation of Optical Properties of Y-Ba-Cu-O Superconductors

Guang-Lin Zhao, Yongnian Xu, W. Y. Ching and K. Wong   63

9 Model Independent Tests of Exitonic Enhancement in High-T_c Superconductors

John P. Ralston   71

10 Bond Asymmetry and High T_c Superconductivity

Daniel C. Mattis and Michael P. Mattis   79

11 Plasmons and High-Temperature Superconductivity in Alloys of Copper Oxides

J. Ruvalds   91

12 Polaron Effects in High-T_c Oxide Superconductors

D. J. Scalapino, R. T. Scalettar and N. E. Bickers   99

13 Electronic Properties of Oxygen Vacancies in La_2CuO_4

Robert V. Kasowski, William Y. Hsu, and Frank Herman   107

14 Model for the Role of Oxygen Defects in Oxygen Defect Superconductors

J. W. Halley and Herbert B. Shore   115

RESONATING VALENCE BOND MODELS

15 The Resonating Valence Bond State and High-T_c
Superconductivity - A Mean Field Theory
G. Baskaran, Z. Zou and P. W. Anderson   129

16 Mean-Field Theory of High-T_c Superconductivity
Andrei Ruckenstein, Peter J. Hirschfeld and J. Appel   137

17 The Large-N Limit of the Hubbard Model: a New
Mean Field Theory for High T_c Superconductors
Ian Affleck and J. Brad Marston   147

18 Topology of the Resonating-Valence Bond State:
Solitons and High T_c Superconductivity
Steven A. Kivelson, Daniel Rokhsar and James P. Sethna   155

19 Neutral Fermion, Charge e Boson Excitations in
the RVB State and Superconductivity in La_2CuO_4-
Based Compounds
Z. Zou and P. W. Anderson   163

20 "Normal" Tunneling and "Normal" Transport: Di-
agnostics for the RVB State
P. W. Anderson and Z. Zou   173

21 Numerical Studies of Superconductivity in the two
Dimensional Hubbard Model
Robert Joynt   183

22 Spin-Gap and Symmetry-Breaking in CuO_2 Layers
and Other Antiferromagnets
Ian Affleck   193

HUBBARD-LIKE MODELS

EXPLICITLY INCLUDING OXYGEN

23   Charge Transfer Resonances and Superconductive
      Pairing in the New Oxide Metals
          C. M. Varma, S. Schmitt-Rink and Elihu Abrahams   211

24   Theory of High T_c Superconductivity in Oxides
          V. J. Emery   227

25   Fermi Liquid Theory of $La_{2-x}Sr_xCuO_4$: Optical
      Properties
          P. A. Lee, G.Kotliar and N. Read   235

26   Pairing Mechanism in Oxide Superconductors: In-
      sights from Small Systems Calculations
          J. E. Hirsch   241

SUBJECT INDEX   253

BACKGROUND

J. W. Halley
School of Physics and Astronomy
University of Minnesota
Minneapolis, Minnesota 55455

Introduction

This volume of reprints and original papers on the theory of high temperature superconductivity has been assembled at a time (fall 1987) of rapid change in the field. No consensus exists concerning the correct basis for a theory which correctly describes the new high temperature oxide superconductors. Some of the flavor of the time is reflected in John Bardeen's historically informative introduction which follows. Despite the circumstances, we believe that this collection has a useful role to play in advancing the theoretical study of high temperature superconductors by assembling ideas and calculations of many of the leading workers and most of the leading points of view for analysis and comparison. Not all of the theoretical uncertainty in this field arises from the rapidly changing experimental picture (reviewed by Allen Goldman below). Some of it arises from failures of communication between theorists and theoretical schools of thought. Thus direct juxtaposition in a coherent format of contrasting approachs (sometimes leading to doubt that the theories can possibly refer to the same material!) can help to clarify matters.

We have organized the papers as follows: After the aforementioned introduction and experimental review, we present a section entitled "BCS-like Theories" which contains papers taking a relatively conservative view and giving or suggesting models which are similar in form to BCS theory of conventional superconductors. This first set of papers begins with one by John Bardeen. Other work described in this section includes a phenomenological analysis uncommitted to mechanism

(Tešanović), and papers suggesting exciton (Freeman, *et al.*, Ralston), phonon (Mattis), plasmon (Ruvalds), bipolaron (Scalapino) and oxygen-vacancy (Halley, Shore) mechanisms of superconductivity. The third section is devoted to work on the resonating valence bond (RVB) model. We begin with the Zou, Baskaran, Anderson paper on a mean field theory of the RVB state of the Hubbard model and related work by Ruckenstein and coworkers. This is followed by a more rigorous approach to mean field theory by Affleck and Marston. We then present a paper by Kivelson, Rokhsar and Sethna on the nature of excitations in the RVB model followed by two papers by Zou and Anderson on the consequences of such an excitation spectrum and another by Affleck on the excitation spectrum. Finally in this section we have a new paper by Robert Joynt on recent numerical results on the Hubbard model. In the fourth section, we turn to models which, as in RVB theories, model the system using a Hubbard-like model, but which take explicit account of the existence of holes on the oxygen as well as the copper sites. Here we include work by V. Emery, C. Varma and coworkers, and P. Lee and coworkers as well as new numerical results by Jorge Hirsch.

The value of this volume will arise mainly from the resolution of the many contradictions in it as they are worked out by the people who study these papers. As regards these contradictions, a striking feature of sections three and four is that the numerical work by Joynt at the end of section three is interpreted to mean that the Hubbard model can lead to superconductivity while the numerical work of Hirsch at the end of the fourth section is interpreted to draw the opposite conclusion. The fact that Hirsch is working at finite temperature while Joynt's results are at $T = 0$ may account for the discrepancy, but this is clearly an important question requiring definitive resolution. More generally, a broad brush comparison of section two with section three shows that the "conservatives" often have a better command of the details of the properties of these materials while the RVB proponents are more sophisticated many body theorists. It is very likely that a correct theory will require a command of both aspects and the work in section IV may be regarded as constituting initial attempts in this direction. A related point is that the short coherence lengths and approximate two dimensional character in the oxide superconductors imply that mean field theory is an especially poor guide to the physics of these systems, a point taken more to heart by the RVB theorists than by the "conservatives" represented in section two. For example, to realize a model along the lines suggested by John Bardeen's remarks concerning strong correlations on the "chains" stabilized by weaker correlations in the "planes" of "1 2 3" clearly requires a BCS-like theory which is far from mean field theory. Spatial disorder is also curiously absent from most of the papers here and in the literature. Experiment as well as estimates of the density of twin boundaries and oxygen vacancies strongly suggest that spatial disorder plays a major role in the new materials. Finally, though many theorists here and elsewhere cite experiment early and often in defense of their theories, suggestions for definitive experiments to distinguish one proposed mechanism or model from another are usually lacking. It is easier to say this than to do something about it. Nevertheless this is an important need.

This volume was assembled as a result of a workshop on Theories of High Temperature Superconductivity held at the Riverwood Conference Center under the auspices and financial sponsorship of the Theoretical Physics Institute at the University of Minnesota. We are grateful to the participants in that workshop, to the Institute and its administrative assistant Shirley Ward and to George Swan for retyping the manuscripts.

J. W. Halley
Minneapolis, January 1988

John Bardeen
Department of Physics
University of Illinois at Urbana-Champaign
1110 West Green Street
Urbana, Illinois 61801

Historical Introduction†

One of the most exciting developments in science in recent times is the discovery during the past year of high-temperature superconducting oxides. It has captured the imagination of the public with front page stories in newspapers such as the *New York Times* and *The Wall Street Journal* and cover stories in magazines like *Time* and *Newsweek*. A superconductor carries a current without resistance. A current flowing around a superconducting ring will flow indefinitely as long as the temperature is maintained below that for the transition from the normal to the superconducting state. Superconductors are now used for a wide variety of applications, from very sensitive detectors of magnetic fields and electromagnetic radiation to high-field superconducting magnets for magnetic resonance imaging tomography and particle accelerators. The largest installation in the world is at the Fermi National Accelerator Laboratory west of Chicago where there is a 4 mile ring of superconducting magnetics. Elementary particles, protons accelerated to very high energies, are confined to flow around the ring by the magnetic fields produced by the magnets.

Applications of superconductivity so far have been limited because of the difficulty and cost of cooling to very low temperatures. Liquid helium required for

†Excerpts from the Van Vleck Memorial Lecture "Theories of High Temperature Superconductivity" delivered at the University of Minnesota, October 1987.

cooling boils at 4 Kelvin or 4 above the absolute zero. This is about 269 below zero on the Celsius scale.

Superconductivity was discovered in 1911, a little over seventy-five years ago, by Kamerlingh Omnes, at the University of Leiden in Holland. He found that the resistance of a rod of frozen Hg dropped to zero when cooled to the temperature of liquid He.

It was soon found that many other elements, alloys and intermetallic compounds become superconducting when cooled to sufficiently low temperatures. Over the years the highest transition temperature had been gradually increased from the 4 K of Hg to 23 K in the compound Nb_3Ge, discovered by scientists at Westinghouse about 15 years ago. This remained the record until about a year ago, when two scientists, Müller and Bednorz, working at the IBM laboratory in Zurich, reported possible superconductivity in a mixture of La and Ba copper oxides at temperatures as high as 30 K, although true superconductivity, zero resistance, was found only below about 15 K. It has been announced recently that they will receive the 1987 Nobel Prize in Physics for their discovery.

Their work started the present explosion of interest, with compounds being discovered with transition temperatures first in the range of 30–40 K, then another class in the range 90–100 K. There have been indications of superconductivity as high as 240 K, or about −10 F, a cold day in the winter in Minneapolis. Very recently there have been reports of room temperature superconductivity by scientists at Georgia Tech. A joke went around at an exciting all-night session last March of the American Physical Society. In the future when you go to the hardware store for a piece of wire, the clerk might ask, "Do you want it normal or superconducting?"

The new superconductors are ceramic oxides, not metals, and have the mechanical properties of ceramics. They are brittle, not ductile like metals. Further, the conduction is highly anisotropic. High current densities are obtained only when the flow is along two-dimensional sheets of copper oxide. The sheets are separated by the divalent barium and trivalent lanthanum. Limiting current densities are much lower when flow is perpendicular to the sheets. High current densities are *not* found in the usual polycrystalline material in which the crystallites have random orientation. Real innovation in materials technology will be required to make the oxides practical for most potential applications.

The first publication of Bednorz and Müller in the German journal, *Zeitschrift für Physic*, in September, 1986, did not create a great deal of excitement and apparently it was not read by very many people. A couple of years ago, Müller, relieved of some administrative obligations, decided to go into a new area of research. About ten years ago superconducting oxides had been discovered that had abnormally high transition temperatures in spite of a rather low density of electrons free to conduct electricity and had a correspondingly low electrical conductivity. Müller decided to look for other possible oxide superconductors with higher transition temperatures. He enlisted the aid of a younger colleague, Bednorz, who largely bootlegged his work, doing it in addition to his assigned tasks. Early work produced no success. Then they heard about a new class of ternary oxides of Ba, La and Cu made by French chemists and decided to give those a try. Initial tests were very promising but not really conclusive. In the title of their first paper were the words "Possible Superconductivity." A Japanese group at the University of Tokyo, that was already doing research on high-temperature superconductivity, began looking into the new oxides, but work was slow to start elsewhere.

A Chinese friend, Chien-hua Tsai, whom I met on a visit to Nanking in 1980 sent to me an interesting letter about how research on the new oxides started in China. The story is a commentary on how few people read the literature these days. Tsai was visiting in Australia at the time the journal with the Müller-Bednorz paper arrived and he read the article. He wrote to a friend at the Institute of Physics in Peking who he knew was interested in high-temperature superconductivity. The friend wrote to Paul Chu at the University of Houston in November and called his attention to the paper that had been lying unread in the library since September. Chu and associates immediately began studying the new oxides. It was after a meeting at MIT in early December, where both Chu and the Japanese scientists reported on their preliminary work, that physicists in this country became generally aware of the exciting results. It was this meeting that initiated the explosion of interest in this country and the rest of the world.

Chu's group not only isolated the structure of a class of compounds with superconducting transitions around 30–40 K, but an associate and former student, M. K. Wu, at the University of Alabama, discovered in January a new class of compounds with transition temperatures in the range 90–100 K.

The first public announcement of the new oxide superconductors was a story in the *People's Daily*, published in Peking. It appeared a few days before a front page story in the *New York Times* at the end of December, 1986. Since then the explosion of interest and activity in the field has been astounding. The only thing I can think of that is comparable is the activity in x-rays following Röntgen's discovery in 1898. More than a thousand papers on x-rays were published in the year following the discovery with a far smaller scientific community than exists today. Then, as now, interest spread to the public as well as the technical press.

Superconductivity is a macroscopic quantum phenomena. Quantum effects are exhibited on a macroscopic scale rather than the scale of atoms and molecules. There was no possibility of explaining the phenomenon before the advent of quantum mechanics in the mid-twenties. Quantum theory, in which matter has both particle and wave aspects, discovered by Heisenberg and Schrödinger in 1924 and 1925, was soon applied to transport of electricity by electrons in metals. In metals, the outer or valence electrons become detached from the atoms that make up the metal and are free to move through the crystal and conduct electricity. Most of the difficulties of the earlier classical Drude theory were resolved in papers published between 1927 and 1932, but superconductivity remained a mystery.

In 1950 it was shown that the transition temperature to the superconducting state depends on the isotopic mass of atoms that make up the metal. This suggested that superconductivity involves an interaction between the conduction electrons and the vibrational motion of the ions in the metal. Even with this clue, it took another seven years before a satisfactory theory was developed by Leon Cooper,

Robert Schrieffer and me at the University of Illinois. It is generally called the BCS theory after the initials of the authors.

This theory was very successful not only in explaining what was known about superconductivity but in predicting new phenomena that later were confirmed by laboratory experiments. The theory is based on a coherent pairing of electrons such that all pairs have identically the same momenta. The pairing results from a long range attraction between electrons.

An analogy has been given to a crowd leaving a football game. A normal metal corresponds to the usual crowd in which people stop and start as opportunity arises. A superconductor would be analogous to people falling into ranks and marching in unison. If an obstacle like a goal post presents itself, people would scurry around and reform ranks on the other side with no loss of momentum. To complete the analogy, people form ranks and break ranks in pairs, like couples of the opposite sex.

In free space, electrons have the same charge and repel one another. To have an attraction that gives rise to the pairing, it is necessary to put the electrons in an environment such that at long distances there is an attractive force that dominates the Coulomb repulsion. In a superconducting metal, the force comes from motion of the ions that make up the crystal lattice in which the electron moves. One can think of one electron displacing a positive ion from its equilibrium position. A second electron feels the force of the displaced ion and is attracted to the same vicinity as the first. The attractive force between the two electrons is mediated by the force between the electrons and vibrational motion of the ions, called the electron-phonon interaction, where the word phonon represents a quanta of vibrational energy of ion motion.

Over the years there have been many attempts to find compounds that become superconducting at higher temperatures. For seventy years, the maximum transition temperature was gradually increased from the 4 K of Hg in 1911 to 23 K in Nb_3Ge in 1972, which remained the record until the discovery of the new superconducting oxides during the past year. Metallic oxides were first studied in the mid-seventies. Two groups of oxides were found that had transition temperatures around 13 K, remarkably high for the small concentration of electrons free to transport electricity. This suggested to Müller and Bednorz that they try to explore metallic oxides in the hope of finding higher transition temperatures.

The explosion of research initiated by Bednorz and Müller has led to the discovery of two new classes of high T_c oxides, one with transitions in the range 30–40 K and one by Chu, Wu and associates in the range 90–100 K, with indications at much higher temperatures. The crystal structures have the common features of sheets of copper oxide between which there are divalent and trivalent ions. The conductivity in the normal state is highly anisotropic, with much higher conductivity in the plane of the sheets than in the transverse direction.

As mentioned earlier, the superconducting properties are also highly anisotropic, with much higher critical currents in the plane than in the transverse direction. This limits the critical currents in the usual polycrystalline material, with crystallites oriented in random directions, to values of the order of 10^3/amps/cm^2, values too small for most potential applications. Values 10 to 100 times larger are required for high-field superconducting magnets. Large current densities are found only when the crystallites are oriented so that flow is in the copper oxide sheets. Thus difficult materials problems must be overcome before many of the potential applications can become a reality.

A possible early application of high temperature superconductivity that has been suggested is for interconnects between silicon integrated circuits. Signals are propagated at the speed of light along superconducting lines.

Regardless of possible applications, the new superconducting oxides are a remarkable scientific discovery. The new oxides have considerably higher transition temperatures than metals with the same concentration of conduction electrons. There are about as many different explanations for the attractive force that gives pairing as there are theorists who have thought about the problem, and that is quite a large number. With more complete experimental data on single crystals becoming available, it should be possible to gradually weed these out and arrive at the correct explanation.

There is considerable controversy as to what gives the attractive interaction in the new high temperature superconducting oxides. In addition to the electron-phonon interaction, a number of other possible ways by which the attractive force is mediated have been suggested. The workshop which was held at the Riverwood Conference Center in Minnesota and on which the present volume is based, discussed the various theoretical proposals that have been made and ways in which the correct explanation can be found. With measurements on single crystals becoming available, it should be possible to narrow down the range of possibilities and eventually arrive at the correct explanation.

The field of high-temperature superconductivity is in a very dynamic research phase with many hundreds of scientists throughout the world frantically searching for the next breakthrough in performance or understanding. With so many competing groups there is undoubtedly much duplication of effort, but with rapid exchange of information through papers and conferences, unnecessary duplication is kept to a minimum. Not enough is known about what the future will bring to have a more directed program. New unexpected discoveries occur frequently. It is a very exciting time for those involved.

Allen Goldman
School of Physics and Astronomy
University of Minnesota
Minneapolis, Minnesota 55455

Review of Experiments

The discovery of extraordinarily high superconducting transition temperatures, well in excess of liquid nitrogen temperatures, in perovskite compounds is one of the great surprises in condensed matter physics. These materials and the prospect for future discoveries have kindled wide interest in superconductivity and superconducting materials research. The central features of superconductivity of these materials are reviewed along with the results of selected measurements. Particular attention is paid to certain critical studies which might be relevant to the mechanism. The issue of the role of sample quality is also discussed. The enormous volume of work in the field prohibits a complete review of the field in a short space despite the fact that the seminal discoveries were made not much more than a year ago.

INTRODUCTION

The perovskites superconductors have the highest critical temperatures of any materials studied to date.[1,2] Depending on the particular extrapolation they may also have enormous critical magnetic fields as well.[3] Whether large critical currents can be achieved in a magnetic field is really not known. The potential impact on technology of ultra-high temperature superconductivity has been the driving force behind the great volume of work in the field over the last year. The frenzied pace of research is consistent with the words of Ginzburg[4] to the effect that the problem of high temperature (ultimately room temperature) superconductivity may be the second most important problem in physical science, behind that of controlled fusion, in terms of its potential impact on society.

Although oxide superconductors with low carrier concentrations, and with transition temperatures above 10 K have been known for a number of years, the new materials were striking in that their transition temperatures are higher than any that might have been expected, given the rate of progress, until well into the middle of the next century., Fig. 1 which contains a plot of transition temperature *vs.* time shows the remarkable change in the field over a very short time.[5]

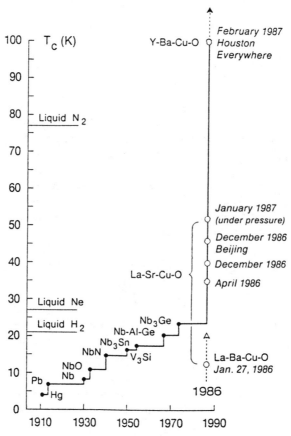

FIGURE 1 Evolution of the superconductive transition temperature subsequent to the discovery of the phenomenon (from Ref. 5).

It is convenient to divide the oxide superconductors into three classes. Class O materials are compounds such as $LiTi_2O_4$ [6] which has a transition temperature of 13.7 K, and $BaPb_{1-x}Bi_xO_3$ [7] which has a transition temperature of 13 K. These materials, although they do not possess remarkably high values of T_c are very puzzling in that their transition temperatures are high despite their relatively low carrier concentrations. The Class 1 materials are those related to La-Ba-Cu-O, which was first reported to be superconducting by Bednorz and Muller.[1] See Fig. 2 for their original data. This material and several related compounds with the same K_2NiF_4 crystal structure have transition temperatures which apparently do not exceed 40 K. The Class 2 materials exhibit superconductivity above liquid nitrogen temperatures, in the range of 90 K, and have the chemical formula $MBa_2Cu_3O_{7-x}$, where the metal M is Y or any of the rare earths except Ce, Pr or Tb.[2,8]

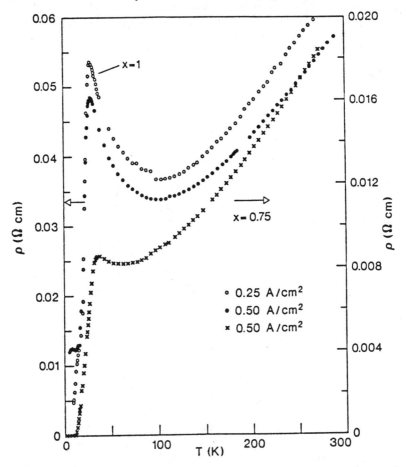

FIGURE 2 Temperature dependence of resistivity in $Ba_xLa_{5-x}Cu_5O_{5(3-y)}$ for samples with $x = 1$ (upper curves, left scale) and $x = 0.75$ (lower curve, right scale) (x nominal). The first two cases also show the influence of current density (from Ref. 1).

It is important to note that *all* of the above mentioned materials exhibit all of the classic features of superconductivity to a significant degree, *i.e.*, zero electrical resistance and the Meissner-Ochsenfeld Effect. In no instance, at this writing is there a confirmed example of superconductivity at temperatures exceeding the order of 95 K, although the scientific literature is filled with numerous reports of anomalous behavior of the resistivity or some other property which is attributed to superconductivity at extraordinarily high temperatures, in some instances in excess of room temperatures. It is inappropriate in this review to discuss this subject at any length, but it may be useful to remark that in considering the possible superconductivity of such systems it is essential that all of the standard criteria be satisfied, and if they are not, and if *all* that is observed is a drop in the resistivity, or even a slight signature of diamagnetism, then the authors are far from demonstrating superconductivity. What is frequently not appreciated is that even a dramatic drop in resistivity may be the signature of a phenomenon other than superconductivity.

Fig. 3 illustrates the spectacular changes in electrical resistivity of materials

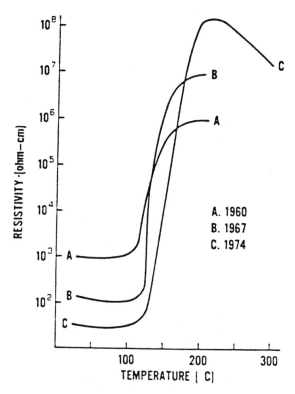

FIGURE 3 PTC characteristics of (A) an early, (B) an intermediate and (C) a modern material (Ref. 9).

which exhibit the so-called PTC effect or positive temperature coefficient of electrical resistivity.[9] This phenomenon is observed in certain semiconducting titanate ceramics in the vicinity of the ferroelectric Curie temperature. The discovery emerged from investigations in the 1950's of controlled valency semiconductors aimed at the development of new thermistor compositions. Although resistivity may change by seven or eight orders of magnitude, the evidence is that the mechanism is a grain-boundary rather than a bulk phenomenon. The phenomenon is directly related to the ferroelectric phase transition in semiconducting barium titanate and its solid solutions with strontium and lead titanates. Semiconductive properties are obtained by substituting trivalent donors (*e.g.* La) on the barium lattice sites, or pentavalent donors on the titanium sites. Possible dopants include Y, most of the rare earths through Er, and in addition, Nb, Ta, , Sb, and Bi. The rough similarity of these materials to those reported to be superconducting at high temperature should be noted.

In the following sections a number of important experimental topics relating to the ultra-high T_c superconductors will be considered. These will include crystal structure and oxygen stoichiometry, macroscopic properties, experiments relating to elemental substitution, evidence of anisotropy, evidence of glass-like character, and the current status of electron tunneling investigations. This is not an exhaustive list as there are many other experimental investigations which will be omitted. Some of these will be discussed in other sections of this volume in the context of specific theoretical models which they may support. I make no claims that this is anything but an anecdotal review, and as such, I would like to apologize in advance to the thousands of authors whose work I do not cite.

CRYSTAL STRUCTURE AND OXYGEN STOICHIOMETRY

The Class 1 oxide superconductors possessing transition temperatures with a limiting value of 40 K are the so-called layered perovskite or K_2NiF_4 phase of $La_{2-x}M_xCuO_4$ with $M =$ Ba, Ca, or Sr. The relevant crystal structure is shown in Fig. 4.[10] The distorted oxygen-defect perovskite structure characteristic of the Class 2 oxide superconductors is shown in Fig. 5.[10] A common feature of both of these structures is a two-dimensional copper-oxygen plane which is believed to be relevant to the conduction process.

An important aspect of the Class 2 materials is the matter of oxygen vacancies. Neutron and X-ray diffraction studies indicate the simultaneous presence of both two and one-dimensional features in the copper-oxygen structures. Neutrons, being sensitive to oxygen, are a finer probe of the location of the oxygen atoms. There are sites in the structures which favor oxygen vacancies. The oxygen vacancies on the top and bottom copper oxygen planes on opposite ends of the unit cell are believed to be organized in chains along the **b** axis rather than as planes parallel to the a–b plane. The other copper-oxygen planes are dimpled as is shown in the figure.

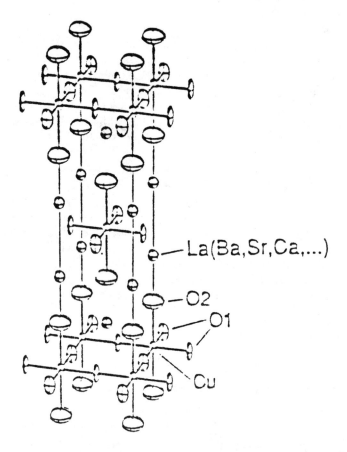

FIGURE 4 Crystal structure of a Class 1 oxide superconductor (from Ref. 10).

These conclusions followed from detailed neutron diffraction studies analyzed using Rietveld refinement techniques.[10]

The significance of the ordering of the oxygen vacancies may be appreciated by noting the transition from an orthorhombic to a tetragonal structure at 950 K. Subsequent cooling of the tetragonal (high temperature phase) through the transition results in a change back into the orthorhombic structure which exhibits high

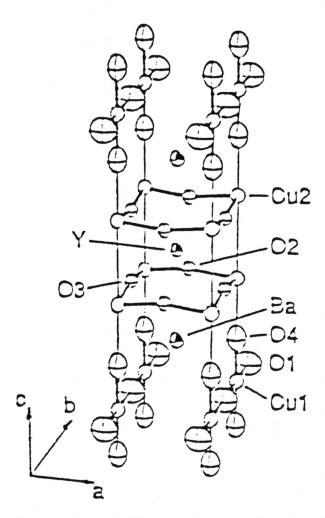

FIGURE 5 Crystal structure of a Class 2 oxide superconductor (from Ref. 10).

temperature superconductivity. On the other hand, rapidly-cooled samples remain tetragonal and are not superconducting at 90 K, but at a much lower temperature. It would appear that 90 K transition temperatures are associated with the linear copper-oxygen chains which are only present in the orthorhombic structure.

Systematic studies of the effect on T_c of oxygen concentration indicate that T_c decreases monotonically with increasing x and it vanishes if the quenched samples remain in the tetragonal phase. An analysis of Cu-O distances suggests that the Cu^{3+} atoms, which are important for many theoretical models, are in the Cu-O chains.[11] At this writing there is no direct evidence of Cu^{3+} atoms in any of the 1–2–3 compounds.

MACROSCOPIC PROPERTIES

The remarkable feature of the superconductivity of the high-T_c materials is that they behave as ordinary superconductors in almost all of their properties, except scaled up by their elevated transition temperatures. The only potential problem is the matter of the critical current density which is extraordinarily low for bulk polycrystalline ceramic materials ($< 10^4 \text{Å/cm}^2$) and can be two to three orders-of magnitude higher for epitaxial films in the **a** and **b** direction.[12] . In addition, many of the properties of conventional superconductors are re-played in the new materials. Thus electron pairs appear to be the carriers of supercurrent as determined in flux quantization and ac Josephson effect studies, and persistent currents are extremely long-lived. Thus, it is tempting to apply the Ginzburg-Landau model to these new materials. Table 1 contains a list of a range of of parameters of $YBa_2Cu_3O_{7-x}$ determined from a detailed analysis of experiments within this context.[13] The important point to note is that the coherence length of the Ginzburg-Landau model is very short implying that behavior other than mean-field behavior, in particular critical phenomena, may be readily observed in the high-T_c materials, in contrast with the case of conventional superconducting materials.

The great similarity of the macroscopic superconducting properties to those of conventional superconductors poses difficulties for critical analysis of theoretical models and suggest that critical experiments will be microscopic rather than macroscopic in character. Thus the matter of carrying out tunneling studies effectively may play a crucial role in the future development of the field. One conventional investigation which is a macroscopic study which may be revealing are the measurements of heat capacity. Studies of bulk polycrystalline samples suggest that the low temperature behavior is linear with temperature.[14] If this variation is indeed associated with the electronic degrees of freedom, then the new high temperature superconductors behave like no other class of superconducting materials in a fundamental manner. The ultimate understanding of this issue depends on the availability of data from high-quality single crystal materials in that disorder, or glass-like two-level systems [15] can also be a source of a linear term in the heat capacity.

The measurement of the Hall Effect is also revealing of the electronic properties of high-T_c materials which may have an ultimate bearing on the understanding of the mechanism.[18] For compounds with the formula $YBa_2Cu_3O_7$ the carrier density corresponds to 1/6 of an itinerant hole per Cu ion at 100 K. The actual carrier concentration is $3.1 \times 10^{21}/\text{cm}^3$ which is rather low. The temperature dependence of the

Hall Coefficient suggests that it is correctly interpreted in terms of carrier concentration. With removal of oxygen from the sample the oxidation state of the samples is reduced, and the carrier concentration is reduced. There may be a transition to a reduced carrier concentration with decreasing oxygen content that corresponds to a change in the average oxidation state of the copper from 2.33+ to 2+. These kinds of considerations may have great relevance for the theory.

Another important issue is the matter of the isotope effect. The most well-known work has involved the replacement of O^{16} by O^{18} in $YBa_2Cu_3O_{7-x}$. The work of the Bell Laboratories group[17] indicates that there is no shift. On the other hand a small shift has been observed in the La-Sr-Cu-O compound which has a lower transition temperature,[18] and even in the 1–2–3 compound a small isotope effect has been reported.[19] The absence of an isotope effect does not preclude a phonon-based mechanism for superconductivity, but it would exclude models that explicitly require one. There have been criticisms of work based on oxygen isotope replacement, rather than on the use of isotopically pure oxygen as a starting material.[21] Experiments of this type are difficult in that it is very difficult to be absolutely certain that processing variations don't produce the changes either observed or not observe in samples with nominally different isotopic constituents. Thus, although the current experimental situation appears to be well-defined, it is possible that the last words have not been written about this very important measurement.

ELEMENTAL SUBSTITUTION

The effects of elemental substitution or chemical doping on the properties of High-T_c superconductors are an important way to elucidate many features of the underlying mechanism for superconductivity. In the few lines available here it is not possible to do justice to the extensive work which has been done on this subject in many laboratories world-wide.

In the case of the 1–2–3 compounds the replacement of Y by rare earths usually results in materials with transition temperatures between 92 and 95 K.[22]. In the case of Yb and Lu the transition temperatures are 87 and 89 K respectively. The exceptions to superconductivity are RE= Ce, Tb and Pr, which are not superconducting. The Pr compound itself is tetragonal rather than orthorhombic. There is apparently no correlation of superconducting transition temperature with the magnetism of the RE ions. There may be some correlation with unit cell volume as has been reported for the La-Sr-Cu-O compounds.

The replacement of Cu by Ni in the 1–2–3 compound produces a dramatic reduction of the superconducting transition temperature which may be an important hint as to the underlying electronic mechanism.[20] A similar effect is observed in the Class 1 oxide superconductors, where the replacement of Cu by Co has also been carried out.

When the Cu atoms in the 1–2–3 compound are replaced by Al there is initially a small down shift in T_c up to 0.1 atomic units at which point there is a drop from 80 K down to zero with increasing Al content.[21] This is important technologically as it explains why thin superconducting films of these compounds are difficult to prepare on Al_2O_3 substrates. A report of the replacement of Cu by Ag, with a reduction of T_c to 50 K also exists.[25]

In any elemental substitution studies great care is needed to ascertain the site substituted and to be certain that the dopant does not go into the material interstitially. Of all of the studies, this type offers great risk of misleading results if the substitution alters other properties of the material in an unknown way. Hence, investigations of this type must be accompanied by very thorough and careful materials characterization efforts.

ANISOTROPY

With the availability of single crystals, it became possible to study anisotropy effects in the high-T_c compounds. Results on the temperature dependence of the electrical resistivity are shown in Fig. 6.[23,24] It is seen that along the **a** and **b** directions the resistivity varies linearly with temperature, whereas along the **c** direction it varies as $1/T$. This result is an important input to any theoretical analysis. Critical currents and critical fields show anisotropies of about a factor of ten or greater.[25] using the Ginzburg-Landau model to interpret critical magnetic field data, it has been inferred that the inter-planar coherence length is about 7 Å, a factor of two larger than the 3.9 Å spacing between Cu-O planes. The in-plane coherence length is about 30 Å. Thus far, no tunneling data has been published on single crystal materials demonstrating any anisotropy of the energy gap.

GLASS-LIKE CHARACTER

The diamagnetism of the zero-field cooled state of the 1–2–3 compound is larger than that obtained when the material is cooled below its transition temperature in a magnetic field. When the magnetic field is turned off, the field cooled material exhibits a time-dependent magnetic moment.[26] Depending upon the temperature and the field, the moment decays very slowly with a functional dependence not exponentially dependent on time. Details of this behavior are very similar to predictions for a superconducting glass.[27] In the latter, the diamagnetic response of superconducting clusters coupled by junctions resembles that of the magnetic response of a spin glass. There is some evidence from the data on superconductors that there is a de Almeida-Thouless line[28] separating metastable from stable regions which would be further evidence of spin-glass-like behavior.

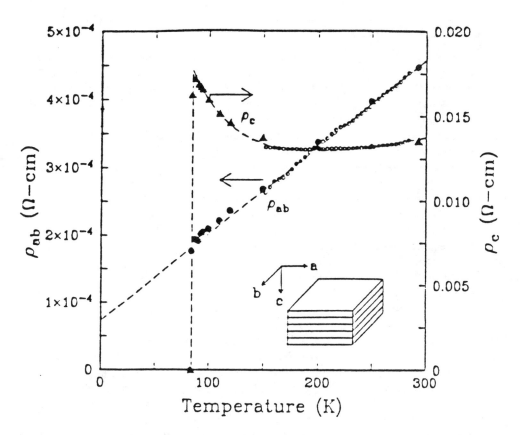

FIGURE 6 Resistivity tensor components parallel to the Cu-O planes (ρ_{ab}) and perpendicular to them, *i.e.*, along the *c*-axis (ρ_c). The small open and large solid symbols are for two different measurements with currents of 500 and 30 mÅ, respectively. The inset schematically illustrates the crystal directions with respect to the Cu-O planes which dominate the conductivity. (from Ref. 24).

It has been argued that the glass-like behavior, which is found in both polycrystalline and single-crystal samples is due to the presence of twin boundaries which are in effect the tunneling junctions needed in the glass picture.[29] If such is indeed the case, and the point can only be tested by carrying out studies on twin boundary free single crystals, then the glass behavior is an interesting phenomenon which has no particular relevance to the mechanism for superconductivity. On the other hand, if the effect were to be found in boundary-free single crystals, and would thus be intrinsic, then an explanation would be required of any proposed mechanism. The existence of twin boundaries may also be intimately connected with the rather limited values of critical currents observed thus far.

ELECTRON TUNNELING

Electron tunneling is potentially an extremely quantitative probe of the super-conducting state of high-T_c compounds. Unfortunately, because of the very short coherence length in these compounds it is essentially only a surface probe, and thus the results are very susceptible to damage at the surface, and lack of stoichiometry. The experimental results, despite the serious problems, have been extensive. Tunneling studies have been carried out on bulk, polycrystalline material, on single crystals and on both single crystal and polycrystalline films.[30] Junctions have been formed by simply depositing a counter electrode on the material, by a break-junction technique, or by using a tip in either a point contact tunneling geometry or with a scanning tunneling microscope. In the case of the latter, it has been very difficult to actually prove that the microscopes are actually operating in a tunneling mode. There is some danger that in the case of any of the measurements in which an electrode is in mechanical contact with the surface of a high temperature superconductor that the tunneling measurement is actually invasive. The pressure exerted by a tip on the surface region can be high enough to change the transition temperature and other parameters.

Break junctions have been used to demonstrate the ac Josephson effect and establish electron pairs as the carriers of charge in the superconducting state. The major characteristic of the data is its great variety. Values of the ratio $2\Delta/kT_c$ ranging from less than 3.5 (the BCS value to more than 15 have been reported. Both symmetric and antisymmetric I-V characteristics have been observed, with the sign of the asymmetry varying. Multiple gap structures have been observed in some measurements. A conductance, linear in voltage has been a common feature of the normal state tunneling in many experiments. At this pont it is not possible to separate the "wheat from the chaff," so that the interpretation of the results must be considered very carefully until a general understanding emerges.

In approaching the interpretation of tunneling data it is very important to keep an open mind as there are no certainties that correct tunneling characteristics when they are obtained will look anything like those studied for the ordinary superconductors. An attempt to discard data that doesn't look right based on old criteria may be very dangerous. It is likely that useful tunneling data will not be obtained until tunneling studies are closely integrated with other surface studies.

DISCUSSION

A recurring problem with experimental studies of high-T_c materials has been sample quality. The early experiments, which were confined to polycrystalline ceramic samples were relatively crude, and in many instances the results reflected the structure of the material rather than an intrinsic superconducting property. Chemical impurities and second phases often complicated results which might have provided microscopic information. The single crystal materials, although far better, nevertheless also suffer from inhomogeneities, the twin boundaries. Thus some results cannot yet be viewed as representing intrinsic properties of the materials.

Tunneling, electronic spectroscopies, and optical studies suffer from difficulties associated with surface quality. There may be oxygen loss from surface, and even surface dead layers which make it very hard to take many of the results as quantitative. With the further development of single crystal film growth techniques, combined with surface diagnostic procedures, it may be possible to surmount some of these difficulties. Optical studies, which were not discussed here, such as infrared absorption, may be superior to tunneling in the long run in that their sampling volume is of the order of the wavelength rather than the coherence length which is very short.

TABLE 1 (from Ref. 13)

Quantity	Max. Value	Min. Value
$\xi(0)_\perp [\text{Å}]$	10	36
$\xi(0)_\parallel [\text{Å}]$	37	16
$\Delta [\text{meV}]$	45	18
$\epsilon_f [\text{eV}]$	0.7	0.02
Δ/ϵ_f	1	0.016

REFERENCES

1. J. G. Bednorz and K. A. Müller, *Z. Phys. B*, **64**, 189 (1986).
2. M. K. Wu, *et al.*, *Phys. Rev. Lett.*, **58**, 908 (1987).
3. T. P. Orlando, *et al.*, *Phys. Rev. B*, **35**, 5347 (1987), and T. P. Orlando, *et al.*, *Phys. Rev. B*, **35**, 7249 (1987).
4. V. L. Ginzburg, *Contemporary Physics (GB)*, **9**, 355 (1968).
5. K. A. Müller and J. G. Bednorz, *Science*, **237**, 1133 (1987).
6. D. C. Johnson, H. Prakash, W. H. Zachariasen, and R. Viswanathan, *Mat. Res. Bull.*, **8**, 777 (1973).
7. A. W. Sleight, J. L. Gillson, and F. E. Biersted, *Solid State Commun.*, **17**, 27 (1975).
8. P. H. Hor, *et al.*, *Phys. Rev. Lett.*, **58**, 1891 (1987); D. W. Murphy, *et al.*, *Phys. Rev. Lett.*, **58**, 1888 (1987); F. Hulliger and H. R. Ott, *Z. Phys. B-Condensed Matter*, **67**, 291 (1987); L. Soderholm, *et al.*, *Nature*, **328**, 604 (1987).

9. Bernard M. Kulwicki, "PTC materials technology, 1955-1980," in *Advances in Ceramics, Vol. 1, "Grain Boundary Phenomena in Electronic Ceramics,"* ed. by L. M. Levinson and D. C. Hill., American Ceramic Society, Inc., Columbus, Ohio, 1981.

10. M. A. Beno, *et al., Appl. Phys. Lett.*, **51**, 57 (1987); J. D. Jorgensen, in *Proceedings of the XVIIIth International Conference on Low Temperature Physics, Jpn. J. Appl. Phys.* **26** Supplement 26-3, 2017 (1987)

11. W. I. F. David, *et al., Nature*, **327**, 310 (1987).

12. P. Chaudhari, *et al., Phys. Rev. Lett.*, **58**, 2684 (1987); R. B. Laibowitz, R. H. Koch, P. Chadhari, and R J. Gambino, *Phys. Rev. B*, **35**, 8821 (1987).

13. A. Kapitulnik, M. R. Beasley, C. Castellani, and C. D. Castro, *Phys. Rev. B*, to be published.

14. M. E. Reeves, T. A. Friedmann, and D. M. Ginzberg, *Phys. Rev. B*, **35**, 7202 (1987); L. E. Wenger, J. T. Chen, G. W. Hunter, and E. M. Logotheits, *Phys. Rev. B*, **35**, 7213 (1987).

15. For an elementary discussion of two-level systems see: Charles Kittel, *Introduction to Solid State Physics*, Sixth Edition, John Wiley and Sons, New York, 1986.

16. N. P. Ong, *et al.*, in *Novel Superconductivity*, eds., S. A. Wolf and V. Z. Kresin, Plenum Press, New York and London, 1987, p. 1061.

17. B. Batlogg, *et al., Phys. Rev. Lett.*, **58**, 2333 (1987); L. C. Bourne, *et al., Phys. Rev. Lett.*, **58**, 2337 (1987).

18. M. L. Cohen, *et al.*, in *Novel Superconductivity*, eds., S. A. Wolf and V. Z. Kresin, Plenum Press, New York and London, 1987, p. 733.

19. Hans-Conrad zur Loye, *et al., Science*, **238**, 1558 (1987); Keven J. Leary, *et al., Phys. Rev. Lett.*, **59**, 1236 (1987).

20. J. M. Tarascon, *et al., Phys. Rev. B*, **36**, 8393 (1987).

21. T. Siegrist, *et al., Phys. Rev. B*, **36**, 8365 (1987).

22. L. Testardi, private communication.

23. T. R. Dinger, *et al., Phys. Rev. Lett.*, **58**, 2687 (1987).

24. S. W. Tozer, *et al., Phys. Rev. Lett.*, **59**, 1768 (1987).

25. T. K. Worthington, W. J. Gallagher, and T. R. Dinger, *Phys. Rev. Lett.*, **59**, 1160 (1987).

26. K. A. Müller, M. Takashige, and J. G. Bednorz, *Phys. Rev. Lett.*, **58**, 1143 (1987).

27. W. Y. Shih, C. Ebner, and D. Stroud, *Phys. Rev. B*, **30**, 134 (1984).

28. J. R. L. de Almeida and D. J. Thouless, *J. Phys. A*, **11**, 983 (1987).

29. G. Deutscher and K. A. Müller, *Phys. Rev. Lett.*, **59**, 1745 (1987).

30. K. E. Gray, M. E. Hawley, and E. R. Moog, in *Novel Superconductivity*, eds., S. A. Wolf and V. Z. Kresin, Plenum Press, New York and London, 1987, p. 611.

BCS-LIKE THEORIES

John Bardeen
Department of Physics
University of Illinois at Urbana-Champaign
1110 West Green Street
Urbana, Illinois 61801

Theories of High-T_c Superconductors[†]

Many theories have been given to account for the high transition temperatures in the oxide superconductors. While most are based on pairing, they differ as to the origin of the attractive interaction that gives rise to the pairs and whether the pairing is weak or strong. If weak ($\lambda < \sim 0.5$), the energy range of the pairing interaction must be wider than is consistent with one mediated solely by phonons. With measurements on single crystals becoming available, giving data covering a wide range of frequencies and temperatures, it is possible to narrow down the possible options. The data show that anisotropic 3D models are required. Observed isotope shifts indicate that phonons must play a role, but additional mechanisms are likely necessary to account for the high T_c's. Thermal, magnetic and transport data are consistent with anisotropic Ginzburg-Landau theory near T_c, with expected departures at low temperatures.

Experiments carried out the past few months have helped clarify the nature of superconductivity in the high T_c oxides. These include measurements of the isotope

[†]Reprinted from *Mat. Res. Soc. Symposium Proceedings* 90, *High Temperature superconductors(1988)*

effect, nuclear spin relaxation and specific heat as well as measurements of magnetic and transport properties on single crystals.

The isotope effect in the $La_{2-x}Ba_xCuO_4$ class of compounds[1] is sufficiently large to indicate that phonons play a major if not the sole role in mediating the effective attractive interactions between electrons. The value of α, for $T_c \sim M^{-\alpha}$, is of the order of 0.25 or larger for ^{18}O replacing ^{16}O. The shift[2] in T_c in $YBa_2Cu_3O_{7-x}$ is of the order of 0.3–0.5 K as compared with 5 K expected if $\alpha = 0.5$ for $T_c = 90$ K. Thus the isotope effect is reduced by a factor of about 5 in going from the LaBaCuO compounds to the YBaCuO compounds. This suggests that some non-phonon mechanism is responsible for the high T_c's in the latter.

Many suggestions have been made for the bosons other than phonons that are responsible for the attractive interaction between electrons.[3] These included excitons (a term used loosely for an electronic charge-transfer excitation such as virtual creation of an electron-hole pair), plasmons, spin-wave fluctuations and others.

Models based on resonating valence bond (RVB) models[4] do not depend on pairing but assume that the Fermi sea is broken down. The elementary excitations are not quasiparticles from a Fermi sea, but are solitons such as charged bosons, or neutral spinless fermions. I shall not discuss such models, but assume that a BCS type pairing is responsible for the superconductivity.

The LaBaCuO compounds have planar sheets of CuO between which there are divalent and trivalent ions. The YBaCuO compounds have two planar sheets of CuO and then a plane which in the tetragonal phase consists of 1D chains of CuO. The two different types of layers are designated loosely as chains (Cu(1)) and planes (Cu(2)). The conductivity in the direction parallel to the planes is about two orders of magnitude larger than in the direction perpendicular to the planes.[5a,b]

A group at Illinois[6] has observed deviations from a BCS-like step at T_c in specific heat of a single crystal of YBaCuO resulting from Gaussian fluctuations of the order parameter. The deviations are consistent with a Landau coherence length of about 10 Å inferred from other data. The amplitude ratio of the fluctuations above and below T_c require a two-component order parameter. This suggests singlet pairing in a BCS theory, but other spins are possible if most of the components are pinned so that they do not fluctuate.

Analysis of magnetic data on single crystals[7] in terms of anisotropic Ginzburg-Landau type models indicate that the coherence distance perpendicular to the planes, $\xi_\perp$, is of the order of 7 Å as compared with $\xi_\parallel \sim 30$ Å in directions parallel to the planes. The short coherence distance perpendicular to the planes suggests that in zero order one should think of the planes and chains as coupled systems with a common transition temperature. Charge transfer matrix elements give Josephson or proximity type coupling from exchange of pairs between planes and chains. When coupled with zero current flow, they have a common phase. The critical current in the perpendicular direction is limited by the coupling energy, as in flow through a Josephson junction.

Warren, et al.,[8] and others have found that there are two distinct resonant frequencies and relaxation times for ^{63}Cu nuclei that they associated with ions in the chains and in the planes. The spins relax by interaction with electrons excited above the superconducting energy gap at the Fermi surface. With decreasing temperature, the number of excited quasiparticles decreases and the relaxation rate drops. The magnitude of the gap can be estimated from the temperature dependence of the relaxation rate.

Since the resonant frequencies are well separated, the two relaxation rates can be measured separately. There are twice as many CuO ions on the planes as on the chains. From the intensity of the signals, Warren, *et al.*, associate an energy gap of about $2.4\,k_B T_c$ with the planes and a gap of about $8\,k_B T_c$ for the chains. This assignment is controversial, with other groups[9] associating the large gap with the planes. Experiments are underway on single crystals that should remove any doubt about the interpretation.

If the large gap is in the chains, it is possible to construct a microscopic theory in which only a phonon mechanism is responsible for superconductivity in the planes but a non-phonon mechanism would be required for the chains. Because of fluctuations, the chains themselves would not be superconducting at finite temperatures, but could be as part of a larger system of chains and planes with a common transition temperature, T_c. The value of T_c is determined by the requirement that the total entropy of the chains and planes is equal to that in the normal state at the second-order phase transition.

To get a semi-quantitative estimate of T_c with a minimum of mathematics, I use the two-fluid model in which the entropy is proportional to T^3 in the superconducting state and to T in the normal state. If $T_1 \sim 2T_c$ is the transition temperature of the chains alone (disregarding localization energies and fluctuations) and $T_2 \sim T_c/2$ that for the planes alone:

$$S_1 = a_1 T^3, \qquad S_{1n} = \gamma_1 T, \qquad \gamma_1 = a_1 T_1^2. \tag{1a}$$
$$S_2 = a_2 T^3, \qquad S_{2n} = \gamma_2 T, \qquad \gamma_2 = a_2 T_2^2. \tag{1b}$$

For the combined system

$$S = (a_1 + a_2)T^3, \qquad S_n = \left(a_1 T_1^2 + a_2 T_2^2\right)T \tag{2}$$

The value of T_c is found from equating S and S_n at the second-order transition:

$$T_c^2 = \beta T_1^2 + (1 - \beta)T_2^2, \tag{3}$$

where $\beta = a_1/(a_1 + a_2)$. For $\beta = 0.2$ and $T_1 = 4T_2$, $T_c = T_1/2 = 2T_2$. These values are consistent with $T_2 \sim 45\,\text{K}$ for the planes alone, $T_c = 90\,\text{K}$ and $T_1 \sim 180\,\text{K}$ for the chains alone.

If the isotope effect is only in the planes and the much stronger superconductivity in the chains is due to some non-phonon interaction, the isotope effect on T_c is given by

$$\frac{\delta T_c}{T_c} = (1 - \beta)\frac{T_2^2}{T_c^2}\frac{\delta T_2}{T_2} = 0.2\frac{\delta T_c}{T_2}. \tag{4}$$

Thus if $\delta T_2/T_2$ corresponds to $\alpha = 0.25$, then $\delta T_c/T_c$, about one-fifth of $\delta T_2/T_2$, would correspond to $\alpha = 0.05$, in agreement with the experiment.[2]

Except for the isotope effect, a similar argument would apply if the roles of the planes and chains are reversed so that the large gap is on the planes. The energy gain from the coupling and making the chains superconducting would compensate for the decrease in transition temperature in the planes from $T_c \sim 2T_c$ to T_c. It would be too costly in energy to localize the electrons on the planes. We will have to wait for the experiments to decide which interpretation is correct.

While a non-phonon interaction is likely required to account for the high transition temperatures in the YBaCuO compounds, a phonon mechanism is not completely ruled out.[10] An upper estimate of $k_B T_c \sim 0.2\,\hbar\omega_{ph}$ has been given,[11] which is barely sufficient for $T_c \sim 100\,\mathrm{K}$. Substantially larger values of T_c, well above $100\,\mathrm{K}$, such as those reported by Bhargava, Herko and Osborne,[12] would indicate that something more than phonons is required.

Experiments are proceeding rapidly so that it should not be too long a wait for some of the answers. As far as I am concerned, it is still a wide-open choice as to what non-phonon mechanism may be involved. I doubt that it will be necessary to go beyond a BCS type pairing to a completely new theory of superconductivity, such as the RVB model, to account for existing data.

ACKNOWLEDGEMENTS

The author is indebted to too many people at Illinois and elsewhere to give proper acknowledgement or to cite all the papers that might be relevant. Discussions with David Pines and C. P. Slichter have been particularly beneficial.

REFERENCES

1. T. A. Faltens, W. K. Ham, S. W. Keller, K. J. Leary, J. N. Michaels, A. M. Stacy, H.-C. zur Loye, D. E. Morris, T. W. Barbee III, L. C. Bourne, M. L. Cohen, S. Hoen, and A. Zettl, *Phys. Rev. Lett.*, **59**, 915 (1987).
2. K. J. Leary, H.-C. zur Loye, S. W. Keller, T. A. Faltens, W. K. Ham, J. N. Michaels, and A. M. Stacy, *Phys. Rev. Lett.*, **59(11)**, 1236–1239 (1987).
3. S. A. Wolf and V. Z. Kresin (editors), *Novel Superconductivity: Proceedings of the International Workshop on Novel Mechanisms of Superconductivity, Berkeley, California*, Plenum Press, New York, 1987.
4. P. W. Anderson, in *Novel Superconductivity: Proceedings of the International Workshop on Novel Mechanisms of Superconductivity, Berkeley, California*,

edited by S. A. Wolf and V. Z. Kresin, Plenum Press, New York, 1987, pp. 295–300.

5. a.) S. W. Tozer, A. K. Kleinsasser, T. Penney, D. Kaiser, and F. Holtzberg, *Phys. Rev. Lett.*, **59(15)**, 1768–1771 (1987);
   b.) S. J. Hagen, T. W. Jing, Z. Z. Wang, J. Horvath, and N. P. Ong, unpublished.

6. S. E. Inderhees, M. B. Salamon, N. Goldenfeld, J. Z. Liu, and G. W. Crabtree, unpublished.

7. T. K. Worthington, W. J. Gallagher, and T. R. Dinger, *Phys. Rev. Lett.*, **59(10)**, 1160–1163 (1987).

8. W. W. Warren, Jr., R. E. Walstedt, G. F. Brennert, G. P. Espinosa, and J. P. Remeika, *Phys. Rev. Lett.*, **59(16)**, 1860–1863 (1987).

9. M. Mali, D. Brinkmann, L. Pauli, J. Roos, H. Zimmerman, and J. Hulliger, *Phys. Lett.*, **A124**, 112–116 (1987).

10. J. C. Phillips, *Phys. Rev. Lett.*, **59(16)**, 1856–1859 (1987).

11. V. L. Ginzburg, "High Temperature Superconductivity: Some Remarks," in *Progress in Low-Temperature Physics*, unpublished; E. G. Maksimov, private communication.

12. R. N. Bhargava, S. P. Herko, and W. N. Osborne, *Phys. Rev. Lett.*, **59(13)**, 1468–1471 (1987).

Zlatko Tešanović
Lyman Laboratory of Physics
Harvard University
Cambridge, Massachusetts 02138

Role of Interlayer Coupling in Oxide Superconductors[†]

The role of interlayer coupling in high-T_c oxide superconductors is considered. Within the assumption that the direct hopping between layers is vanishingly small, we find that the interlayer coupling in the particle-particle channel caused by interactions plays a very important role in the enhancement of T_c and stabilizes the superconducting order with respect to fluctuations. The interlayer coupling in the particle-hole channel plays a relatively minor role. The introduction of a small interlayer hopping matrix element reduces T_c in the lowest-order approximation.

The recent discovery of high-T_c superconductivity in Cu-based oxide superconductors[1-6] has generated enormous interest in the origin of pairing in these materials. Several theories have appeared proposing that a new mechanism for superconductivity may be responsible for $T_c \sim 100$ K. The mechanisms invoked include the resonating valence-bond state,[7-11] antiferromagnetic spin fluctuations,[12] excitons,[13] and plasmon-mediated pairing,[14,15] as well as the bipolarons.[16] The anomalously strong electron-phonon interaction has also been pointed to as a possible source of high T_c.[17-19] Most of the above models, although not all of them, emphasize a

[†]Reprinted from *Phys. Rev. B*, **36**, 2364 (1987).

two-dimensional character of the electronic states in oxide superconductors, with "nesting" of the Fermi surface in a square lattice playing a prominent role in the enhancement of T_c. The band-structure calculations for the layered perovskites La$_{2-x}$Ba$_x$CuO$_4$,[17,18] and YBa$_2$Cu$_3$O$_{6.9}$,[19] reveal the antibonding Cu$_{d_{x^2-y^2}}$ and oxygen p orbital derived bands at the Fermi level with very little dispersion along the z axis, supporting the picture of a quasi-two-dimensional layered character of these compounds.

In this context the question arises as to why the superconducting transition is so sharp, at least in the resistivity,[20] and what is the nature of fluctuations in these systems? In ordinary layered superconducting structures, the phase fluctuations of the order parameter in different layers are stabilized by the hopping matrix element describing electron tunneling between the layers.[21] If the direct hopping between layers is vanishingly small in the conduction band of oxide superconductors, as suggested by band-structure calculations, we have to think of other mechanisms which can prevent phase fluctuations from destroying the long-range order.

I propose in this paper that the interlayer coupling due to *interactions* plays an important role *both* in stabilizing the long-range order in oxide superconductors and providing a mechanism for *further enhancement* of T_c. The "Josephson-like" coupling between the layers which arises through interaction of electrons in the neighboring CuO$_2$ planes enhances the transition temperature within the layer, *irrespective* of its sign. This situation is quite different from that encountered in most ordinary layered compounds.[22] There the layer coupling is due to direct interlayer hopping found in the single-electron band structure. We propose that such direct hopping is only a higher-order effect in oxide superconductors, and that their conduction band is basically two dimensional. The electron transport arises only via scattering to the three-dimensional bands, above and below the Fermi level, arising through Coulomb interaction. This makes oxide superconductors a very special novel class of layered materials.

The Hamiltonian of our model can be written as follows:

$$H = \sum_i H_i^{\text{intra}} + \sum_{\langle i,j \rangle} H_{\langle ij \rangle}^{\text{inter}}, \tag{1}$$

where H_i^{intra} is the layer BCS-type Hamiltonian given by

$$H_i^{\text{intra}} = -2D \sum_{\mathbf{k},\sigma} \gamma_{\mathbf{k}} c_{\mathbf{k}\sigma,i}^{\dagger} c_{\mathbf{k}\sigma,i}$$
$$+ V \sum_{\mathbf{k},\mathbf{k}'} c_{\mathbf{k}\uparrow,i}^{\dagger} c_{-\mathbf{k}\downarrow,i}^{\dagger} c_{\mathbf{k}'\downarrow,i} c_{-\mathbf{k}',\uparrow,i}. \tag{2}$$

In (2) $c_{\mathbf{k}\sigma,i}^{\dagger}$ is the creation operator for electrons in the ith CuO$_2$ layer, with the linear momentum $\mathbf{k}$ within the layer and spin σ, $\gamma = \cos(k_x a) + \cos(k_y a)$, $2D$ is the bandwidth, and $\langle i,j \rangle$ indicates that i and j are neighboring layers. The attractive interlayer interaction $V = -|V|$ is assumed, originating from some of the proposed mechanisms, which we *do not specify* here. The interlayer coupling is contained in $H_{(i,j)}^{\text{inter}}$, given by

$$H_{(i,j)}^{\text{inter}} = -t \sum_{\mathbf{k},\sigma} c_{\mathbf{k}\sigma,i}^{\dagger} c_{\mathbf{k}\sigma,j} + Y \sum_{\mathbf{k},\mathbf{k}',\alpha,\beta} c_{\mathbf{k}\alpha,i}^{\dagger} c_{-\mathbf{k}\beta,j}^{\dagger} c_{\mathbf{k}'\beta,j} c_{-\mathbf{k}',\alpha,i}$$
$$+ W \sum_{\mathbf{k},\mathbf{k}',\alpha,\beta} c_{\mathbf{k}\alpha,i}^{\dagger} c_{-\mathbf{k}\beta,i}^{\dagger} c_{\mathbf{k}'\beta,j} c_{-\mathbf{k}',\alpha,j}. \tag{3}$$

The first term in (3) is the direct hopping between the layers, while the second and the third terms describe the interlayer coupling due to interactions. This is a very general form of the interlayer coupling. As already emphasized, t is very small in oxide superconductors. Y denotes the interlayer coupling constant in the particle-hole channel and may contain contributions from plasmon-, exciton-, and phonon-assisted transitions, as well as direct Coulomb interaction between charged layers. W is the coupling constant in the particle-particle channel, which is not due to any density fluctuations, but can arise through Coulomb interaction causing transitions from the band at the Fermi surface to some of the fully occupied or empty bands away from the Fermi level, with finite dispersion along the z axis. From the calculations of Refs. 17 and 18, it follows that there are a number of such bands within a few eV of the Fermi level. This is a peculiarity of the band structure which we suggest is very important for superconductivity in these materials. Fig. 1 depicts the type of processes contributing to W.

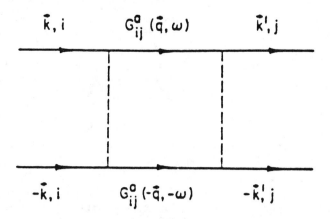

FIGURE 1 Type of interband scattering processes contributing to W. Two electrons in layer i are scattered by the Coulomb interaction into the bands away from the Fermi level. If these bands have a finite dispersion along the z axis, the electrons propagate to a neighboring layer j where they are scattered back to the original band at the Fermi level. $G_{ij}^{\alpha}(q,\omega)$ are the propagators of the intermediate band states. The dotted lines represent Coulomb interactions.

We now assume that the Hamiltonian (1) can be treated within the weak coupling BCS theory. This seems a sensible starting point to obtain some qualitative results, particularly if the pairing interaction is of the electronic origin. We first set $t = 0$. Then the BCS expression for the free energy can be written as

$$F = -\sum_{i,j} \Theta_{ij} \Delta_i^* \Delta_j - \frac{1}{Y} \sum_{i,\gamma} \Gamma_{i\gamma}^* \Gamma_{i\gamma} - T \sum_{\mathbf{k},\omega_n} \sum_{i,j} \ln \det G_{ij}(\mathbf{k}, \omega_n) \qquad (4)$$

where Δ_i and $\Gamma_{i,\gamma=\pm 1}$ are, respectively, the intralayer and interlayer order parameters, arising from the anomalous expectation values of the type $\langle c_{\mathbf{k}\uparrow,i} c_{-\mathbf{k}\downarrow,i} \rangle$ and $\langle c_{\mathbf{k}\uparrow,i} c_{-\mathbf{k}\downarrow,i+\gamma} \rangle$, and ω_n's are the Matsubara frequencies. We have assumed that order parameters are uniform within a layer. The matrix $G_{ij}(\mathbf{k}, \omega_n)$ is given by

$$G_{ij}(\mathbf{k}, \omega_n) = \frac{1}{(\omega_n^2 + \xi_{\mathbf{k}}^2)^{1/2}} \left[\begin{pmatrix} i\omega_n - \xi_{\mathbf{k}} & \Delta_i \\ \Delta_i^* & -i\omega_n - \xi_{\mathbf{k}} \end{pmatrix} \delta_{i,j} + \begin{pmatrix} 0 & \Gamma_{i\gamma} \\ \Gamma_{i\gamma}^* & 0 \end{pmatrix} \delta_{i,j-\gamma} \right],$$
$$(5)$$

where $\xi_{\mathbf{k}} \equiv -2D\gamma_{\mathbf{k}} - \mu$, μ being the chemical potential. The coupling of the order parameters Δ_i's on different layers is determined by

$$\Theta_{ij} = \frac{1}{N} \sum_{k_z} \frac{e^{ik_z(z_i - z_j)}}{V + 2W \cos(k_z d)},$$

where N is the number of layers, d is the separation between them, and $z_i \equiv id$.

Note that expanding the last term in (4) to second order in Δ_i and $\Gamma_{i\gamma}$ *does not* generate any interlayer coupling. To this order, the interlayer coupling is solely due to Θ_{ij}, and it is nonzero only for *finite* (*i.e.*, nonzero) W. Therefore, the finite coupling constant for interlayer interactions in the particle-particle channel generates the "Josephson-like" coupling between order parameters on different layers. If $W = 0$ the order parameters of different layers are uncoupled, and phase fluctuations will prevent the establishment of a true long-range order. The wealth of available experimental information suggests that oxide superconductors behave in a way indicative of a long-range superconducting order.[23] We propose here that in these materials where t seems to be very small (*i.e.*, $t < 0.1$ eV), it is the above mechanism of the interlayer coupling that is primarily responsible for stabilizing the superconducting state. An objection can be raised here that the very type of interband transitions that contribute to W could result in the effective hopping matrix element of a similar magnitude. We can show,[24] assuming short-ranged interactions acting only within a layer, that the contributions to t of interband processes in Fig. 1 either vanish by symmetry or are negligible in comparison with W. This ensures that our model, in which t is initially set to zero, has an internal consistency. Similarly, we do not expect any phonon or low-energy plasmon processes to be effective in the interband scattering; the characteristic energies for these processes are way below the 1 eV scale.

Let us now search for the self-consistent solution of this BCS-like theory by minimizing the free energy in Eq. (4) with respect to Δ_i and $\Gamma_{i\gamma}$. It is easy to convince oneself that the values of the order parameters at the minima of (4) satisfy $\Delta_i \Gamma_{i\gamma} = 0\,\forall i, \gamma$. This is the consequence of $t = 0$. If $Y > 0$, $\Gamma_{i\gamma} = 0$ at any temperature. For $Y < 0$, the phase with nonzero Γ is possible only if $|V| + 2|W| < |Y|$. In view of the above discussion, and the fact that most of the proposed theories emphasize the intralayer coupling as a source of high T_c, this possibility appears unlikely. Therefore we limit ourselves to the case when $|V| + 2|W| > |Y|$. Then there is a single second-order phase transition, to a state with $\Delta_i \neq 0$. Let us first assume that W is negative. Then the minimum of the free energy is obtained for $\Delta_n = |\Delta| \exp(i\phi)$, with the magnitude and the phase of the order parameter identical in all layers. If W is positive, the favored state is $\Delta_n = |\Delta| \exp(i\phi + in\pi)$, with the constant magnitude, but with the phase which changes by $\pm\pi$ as one crosses from one layer to the next. We denote these two states as favored (F) and antifavored (AF), respectively. The transition temperature depends *only* on $|W|$ and can be written as

$$T_c = 1.14\,\omega_0 \exp\left(-\frac{1}{\lambda_V + 2\lambda_W^*}\right), \tag{6}$$

with $\lambda_V \equiv N(0)|V|$, $\lambda_W \equiv N(0)|W|$, $N(0)$ being the density of states at the Fermi level, and

$$\lambda_W^* = \frac{\lambda_W}{1 + 2\lambda_W \ln(\Omega_0/\omega_0)}.$$

It is assumed here that ω_0 is the frequency cutoff of the pairing mechanism responsible for intralayer attraction, while Ω_0 is the energy scale characteristic of processes contributing to W.

The fact that W enhances the transition temperature regardless of its sign has a formal resemblance to the two-band model proposed by Lee and Ihm.[25] In our case it is the interlayer coupling arising through interactions that plays a role similar to the interband scattering mechanism of Ref. 25. It is conceivable that the "excitonic" type of processes shown in Fig. 1 could lead to sizable W. One can very roughly estimate $W \sim n_b \tilde{N}(0) V_c^2 (\tilde{t}^2/K^2)$, where $\tilde{N}(0)$ is the typical density of states for bands near the Fermi level, n_b is the number of bands which contribute to the process, and $\tilde{t}$ and K are characteristic dispersion and the distance from the Fermi level for these bands. V_c measures the strength of the interband Coulomb interactions. While $\tilde{t}$ appears to be small in these materials, there is some evidence that V_c may be quite large, of order of few eV.[26] If we take as typical values $\tilde{t} \sim 0.3\,\mathrm{eV}$, $N(0) \sim \tilde{N}(0) \sim 1.0\,\mathrm{eV}^{-1}$, $V_c \sim 3.0\,\mathrm{eV}$, and $K \sim 3.0\,\mathrm{eV}$ we find $W \sim 0.1 n_b$. As there are several bands near the Fermi surface that could contribute to this scattering, it appears likely that the value of W due to such "excitonic" processes could give a significant contribution to the enhancement of T_c.

What would be the experimental manifestations of the F and AF states? One clearly expects some interesting behavior in the upper critical magnetic field, in the direction parallel to the layers. This upper critical field is the subject of our current work,[24] and one would hope that H_{c2} measurements on a single-crystal sample could be used to differentiate between F and AF types of susperconducting states. There should be no Josephson effect (to the lowest order) between the AF superconductors and the ordinary superconductors if the junction is manufactured in such a way that the plane of contact is perpendicular to the layers of oxide superconductor. But for a careful determination of the stable state in oxide superconductors, it will be first necessary to produce single crystals of sufficient size.

Finally we discuss the effect of small but finite t, in the interlayer Hamiltonian given by Eq. (3). Such matrix elements can arise through phonon- or plasmon-assisted hopping, for example. To evaluate T_c in this case, we assume that coupling constants V and W do not change significantly with the introduction of small dispersion along the z axis. To justify this assumption it would be necessary to invoke a specific assumption about the nature of the interaction leading to superconductivity, and to calculate V and W within a given model. Clearly, this is beyond the scope of this paper. Therefore we proceed, keeping V and W fixed. The calculation of the lowest-order correction in T_c involves expanding the logarithm of the determinant in Eq. 4 in power series in t and finding the first contributing term. In the F state, after some algebra, we find

$$T_c \cong T_{c0} \exp\left(-\frac{t^2}{2\pi^2 T_{c0} D}\right), \tag{7}$$

while, for the AF state one obtains

$$T_c \cong T_{c0} \exp\left(-\frac{7\zeta(3)t^2}{2\pi^2 T_{c0}^2}\right), \tag{8}$$

In the above, T_{c0} is the transition temperature for $t = 0$. Therefore, within the limits of our assumption, introducing the small hopping matrix element between layers reduces T_c in both types of superconducting states. This reduction is much more pronounced for the AF superconductor, as $T_{c0} \ll D$. one should add that the effect of t on the F state is very dependent on assumptions about density of states in the conduction band.

If the intralayer coupling owes its strength to a quasi-two-dimensional nature of electronic states, as is now claimed by a number of authors, we expect that the coupling constants will either not change or will decrease if the hopping between layers is allowed. The results of Eqs. (7) and (8) are consistent with such behavior, and one would generally expect that in such a case an increased direct hopping between layers would be unfavorable for high-T_c superconductivity. One should mention, however, that in Ref. 12 it is proposed that, if the electron-electron attraction is induced by spin fluctuations, making a system more three dimensional would increase the pairing interaction and would also decrease a depairing effect arising from very strong inelastic electron-electron scattering. In such a situation, we cannot make any firm conclusions, since the reduction predicted by (7) and (8) may be offset by an increase of the coupling constants.

Finite t may also lead to an additional second-order phase transition, to the state with $\Gamma \neq 0$. Such a state would be characterized with two different excitation gaps, and would have a very anisotropic excitation spectrum. The anisotropy of the spectrum has been recently used by Maekawa, Ebisawa, and Isawa[27] to explain the discrepancy between the infrared and the tunneling data for the gap parameter.

ACKNOWLEDGEMENTS

I am grateful to B. I. Halperin, G. Baskaran, M. Stephen, and J. Ruvalds for sharing their insights on the oxide superconductors and for useful discussions. This work has been supported by the National Science Foundation through Grant No. DMR85-14638 and the Harvard Materials Research Laboratory.

REFERENCES

1. J. G. Bednorz and K. A. Müller, *Z. Phys. B*, **64**, 189 (1986).
2. C. W. Chu, *et al.*, *Phys. Rev. Lett.*, **58**, 405 (1987).
3. R. J. Cava R. B. van Dover, B. Batlogg, and E. P. Rietman, *Phys. Rev. Lett.*, **58**, 408 (1987).
4. M. K. Wu, *et al.*, *Phys. Rev. Lett.*, **58**, 908 (1987).
5. P. H. Hor, *et al.*, *Phys. Rev. Lett.*, **58**, 911 (1987).
6. J. M. Tarascon, L. H. Greene, W. R. McKinnon, and G. W. Hull, *Phys. Rev. B*, **35**, 7115 (1987).
7. P. W. Anderson, *Science*, **235**, 1196 (1987).
8. G. Baskaran, Z. Zhou, and P. W. Anderson, *Sol. St. Commun.*; reprinted in this volume.
9. S. Kivelson, D. Rokhsar, and J. Sethna, *Phys. Rev. B*, **35**, 8865 (1987).
10. M. Cohen and D. H. Douglass, *Phys. Rev. B.*, **35**, 8720 (1987).
11. H. Fukuyama and K. Yosida, *Jpn. J. Appl. Phys.*, to be published.
12. P. A. Lee and N. Read, *Phys. Rev. Lett.*, **58**, 2691 (1987).
13. C. M. Varma S. Schmitt-Rink, and E. Abrahams, *Proceedings of the International Conference on New Mechanisms of Superconductivity*, ed. by V. Kresin and S. Wolf, Plenum, New York (1987); reprinted in this volume.
14. V. Kresin, *Phys. Rev. B*, **35**, 8716 (1987).
15. J. Ruvalds, *Phys. Rev. B*, **35**, 8869 (1987).
16. P. Prelovšek, T. M. Rice, and F. C. Zhang, *J. Phys. C*, to be published.
17. L. F. Mattheiss, *Phys. Rev. Lett.*, **58**, 1028 (1987).
18. J. Yu, A. J. Freeman, and J.-H. Xu, *Phys. Rev. Lett.*, **58**, 1035 (1987), and the following two papers in this volume.
19. L. F. Mattheiss, unpublished.
20. Many groups now have samples exhibiting the width of a resistivity transition of the order of 1% or less.
21. R. A. Klemm and K. Scharnberg, *Phys. Rev. B*, **24**, 6361 (1981).
22. A good early review can be found in R. A. Klemm, *Ph.D. Thesis*, Harvard University (1974) unpublished.
23. I find it impossible to do justice to the enormous amount of experimental information already available on oxide superconductors by listing only a few

references. Most of the conventional experimental techniques have been applied to these materials and they seem to indicate that the long-range order is well established in the superconducting state.

24. Z. Tešanović, unpublished.
25. D. H. Lee and J. Ihm, unpublished.
26. G. Baskaran, private communication.
27. S. Maekawa, H. Ebisawa, and Y. Isawa, *Jpn. J. Appl. Phys.*, to be published.

A. Freeman, Jaejun Yu, and C. L. Fu
Department of Physics and Astronomy
Northwestern University
Evanston, Illinois 60201

Electronic Structure and Properties of Quasi-Two-Dimensional Layered Superconducting Perovskites: $La_{2-x}M_xCuO_4$ (M = Ba, Sr,...)[†]

Results of highly precise all-electron local density calculations are presented for the quasi-2D superconductors, $La_{2-x}M_xCuO_4$ (M = Ba, Sr, ...) and correlated with a number of observed properties of these materials for varying x. The dominant role of the van Hove saddle point singularity (SPS) on the density of states (DOS) and the DOS derived properties (specific heat, magnetic susceptibility, T_c) is emphasized—as is the fact that this SPS represents the classic Lifshitz transition *vs.* x (and/or pressure) with its possible thermodynamic and galvanomagnetic anomalies. Estimates of the enhancement factor, λ, from a variety of measured specific heat values and our calculated DOS indicate that these materials may possibly be weak coupling superconductors. Crude estimates of coherence lengths ($\xi_{\parallel}$ and $\xi_{\perp}$) and upper critical fields ($H_{c2\parallel}$ and $H_{c2\perp}$) are presented.

The high T_c superconductors[1], $La_{2-x}M_xCuO_4$ (M = divalent metal Ba, Sr, ...) continue to grow in experimental and theoretical interest.[2-4] In a previous paper,[5] we presented some first results of the electronic structure and properties of body centered tetragonal (bct) La_2CuO_4 which showed that the electronic structure and

[†]Reprinted from *Phys. Rev. B*, **36**, 7111 (1987).

properties are dominated by strong in-plane interactions between the Cu d and O(1) $2p$ electrons. The results of this highly precise all-electron local density full potential linearized augmented plane wave[6] (FLAPW) energy band calculation of the band structure, charge densities, Fermi surface, etc., demonstrated: (i) that the material consisted of metallic Cu-O(1) planes separated by insulating (dielectric) La-O(2) planes and (ii) that this 2D character and alternating metal/insulator planes would have as some of their most important consequences, strongly anisotropic (transport, magnetic, etc.) properties.[5] While measurements to date have been done only on powders, it is clear that synthesis and measurements on single crystals of these materials are an urgent priority and results are expected shortly.

In this paper, we present detailed results on the electronic structure, especially density of states, which we correlate with varying composition x of the M = divalent additions and resulting properties. We emphasize that a highly important role is played by a strong 2D van Hove saddle point singularity (SPS) in the electronic band structure and properties for $x > 0$. This SPS dominates the density of states at and near E_F. As reported elsewhere,[7] it also strongly affects the Fermi surface and the generalized susceptibility, $\chi(\vec{q})$. It turns out that this material demonstrates vividly the classic case of an SPS first discussed in the work of Lifshitz[8] who showed how a change in the topology of the Fermi surface is accompanied by the appearance of various anomalies in the thermal and electronic properties. Later, Dagens[9] and others have emphasized the occurrence of phonon anomalies near such a FS topological phase transition and possible changes in T_c as well.

The calculated band structure was shown previously[5] along high symmetry directions in the Brillouin zone. Only flat bands, i.e., almost no dispersion, were found along the c axis, demonstrating that the interactions between the Cu, O(2) and La atoms are quite weak. However, along the basal plane directions there are very strong interactions between the Cu-O(1) atoms leading to large dispersions and a very wide bandwidth ($\sim 9\,$eV). The band structure near E_F has a number of interesting features. What is especially striking is that, in contrast to the complexity of its structure, only a single free electron-like band crosses E_F and gives rise to a simple Fermi surface (cf., Fig. 1). Since this band originates from the Cu $d_{x^2-y^2}$-O(1) $p_\parallel$ orbitals confined within the Cu-O(1) layer, it exhibits clearly all the characteristics of a two dimensional electron system. Particularly striking in Fig. 1 is the occurrence of a van Hove saddle point singularity (note arrow in the figure) near the G_1, N, and G_4 points in the Brillouin zone (BZ). Such an SPS is expected, and found, to contribute strongly, via a singular feature, to the density of states (DOS).

In fact, as shown in Fig. 2 and insert to Fig. 1, the DOS of bct La_2CuO_4 has a peak structure near E_F which arises from saddle points determined by the two-dimensionality of the $d_{x^2-y^2}$-$p_\parallel$ band. Since this band is almost half-filled, the peak in the DOS is located very close to the Fermi energy (about 0.1 eV below E_F). As

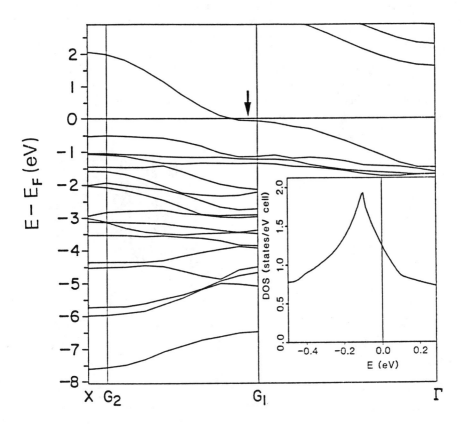

FIGURE 1 Energy band structure of $La_{2-x}M_xCuO_4$ along some high symmetry directions. Note the van Hove singularity labelled by the arrow. The insert shows the van Hove singularity in the DOS.

a result, the DOS at E_F, $N(E_F)$, is expected to be very sensitive to the variation of the chemical potential and hence, as we shall see, to the composition x.

The general features of the valence band structure at and below the E_F formed from a superposition of Cu d, O(1) and O(2) p orbitals, are reflected in the DOS (*cf.*, Fig. 2). The strong interactions of Cu d and O(1) $p_{\parallel}$ in the plane are seen from the large splitting of their projected DOS into two peaks separated by 4 eV from O(1) and Cu. By contrast, the O(2) projected DOS yields only a single broad peak due to the weak Cu d and O(2) p hybridization.

We had seen[5] that the quasi-2D properties of the electronic structure are also supported by plots of the charge densities of electrons at E_F. This charge density consists mainly of Cu $d_{x^2-y^2}$ and O(1) $p_{\parallel}$ hybridized orbitals in the plane with some additional contributuion of the Cu d_{z^2} and O(2) p_z components. There is essentially no electron density around the La site at E_F. This means that the

La atoms do not contribute directly to the dynamical processes involving electrons near E_F. Further, an analysis of the band structure[5] shows that the $5d$ level of La lies more than 1 eV above E_F; the $5p$ levels of La were found[5] to lie far below E_F ($\sim$ 15 eV). Thus, it is a fairly good approximation to consider the La atoms to be described in chemical terms as La^{3+} ions. One important consequence of this is that dilute substitution for La by trivalent magnetic rare-earths, like Pr, Nd, etc., would not affect T_c because there are no conduction electrons which can carry their magnetic pair-breaking interactions[10].

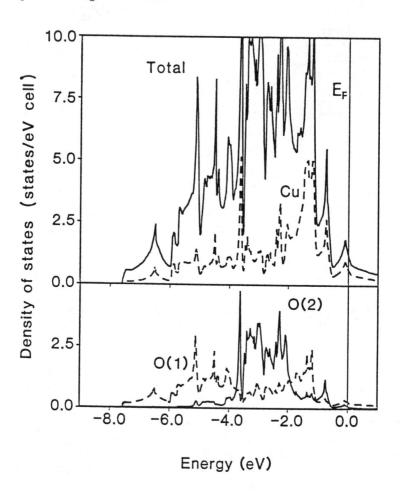

FIGURE 2 Total and projected density of states for the valence bands (essentially all from Cu d - O p electrons). E_F is given for the $x = 0$ composition.

The remarkable two dimensional nature of the electronic structure just described leads to a simple picture of the conductivity confined essentially to the metallic Cu-O(1) planes separated by insulating (ionic) planes of La-O(2). This picture is strongly confirmed by independent calculations[11] which model $La_{2-x}M_xCuO_4$ as a single slab consisting of a Cu-O(1) layer sandwiched by one La-O(2) layer on each side. (Note that such a slab has the correct stoichiometry and is charge neutral.) The electronic structure near E_F is dominated by the same single band of 2D p-d bonding character; the nesting feature (with zone boundary spanning vector) and the van Hove SPS in the DOS are reproduced with this slab approach.

In view of the results presented above, we would expect—as a first approximation—that the introduction of divalent elements (*e.g.*, M = Ba, Sr, etc.) as substitutional replacements for La would not change any major feature of the band stucture, charge density, DOS, etc. Thus, the use of a rigid band approximation to treat the case of alloys, $La_{2-x}M_xCuO_4$, may be considered as a quite good first approximation when x is small (≤ 0.3). (This has been confirmed by independent virtual crystal approximation calculations.[11]) Hence, in this spirit, the variation of composition x in $La_{2-x}M_xCuO_4$ can be taken into account simply as a change in the position of E_F, that is $E_F = E_F(x)$. Further, as stated above, since E_F lies very close to the SPS, $N(E_F)$ is extremely sensitive to the position of E_F relative to the singular point. As a function of x, $N(E_F)$ varies from 1.2 states /eV-cell at $x = 0$ to 1.9 states/eV-cell at $x = 0.16$. This large variation in $N(E_F)$ will immediately affect a number of properties such as the magnetic susceptibility, specific heat, etc. At the critical point, however, the logarithmic singularity is not as divergent in the DOS as in the case of an exact two-dimensional system even though the profile of the DOS looks like a logarithmic shape. This behavior is expected from the fact that there are small residual 3D interactions between the metallic Cu-O(1) layers which exhibit the quasi-2 dimensionality of the system.

As a result of this singular behavior of the DOS, many other properties of these alloys are expected to be determined by the Lifshitz-van Hove singularity. As is well-known, Lifshitz[8] discussed various anomalous thermodynamic and galvanomagnetic properties of a metal that would occur when applied pressure causes E_F to pass through the SPS, *i.e.*, a change of the chemical potential induced by pressure, $\mu = \mu(p)$. In our case, alloying with M_x additions serves to change E_F because of a change in the number of electrons, n, giving us a change $\mu = \mu(n)$ or $\mu = \mu(x)$. Since the van Hove SPS is near E_F for $x = 0$, lowering n (increasing x) leads to a critical composition at which the singular point and $E_F(x)$ coincide. Thus, x could be viewed as playing the role of pressure in Lifshitz's treatment. In addition, we can also introduce pressure as a second variable since, for a given x, pressure could be applied to again drive E_F through the SPS.

The crossing of the SPS by E_F with changing x (or pressure) is expected to have dramatic effects on various properties including the lattice parameters, bulk moduli, elastic constants, Young's modulus, etc.[12] Elsewhere[13] we show that variations of the lattice parameters a and c of the bct structure have already been observed for variation of x which directly reflect the important role played by the existence of the

SPS. One major effect of the van Hove singularity on the properties of the system is the anomalous behaviour of T_c with varying compositon x in $La_{2-x}M_xCuO_4$. With increasing x, the superconducting critical temperature T_c increases[14] rapidly from 14 K for $x \simeq 0.05$ to a maximum of 37 K for $x \simeq 0.2$ but then drops sharply for larger x values. Such anomalies in T_c under pressure have been treated[15] by using a modified BCS equation. As seen above, in contrast to their 3D van Hove singularity[15] which gives $\delta N(E) \alpha |E_F - E|^{1/2}$, the singular behavior of the DOS in our quasi-2D system results in the logarithmic type (*i.e.*, strong) change of the DOS. Thus, we expect that the correction, δT_c, due to the singular part of $N(E)$ will be much larger than that obtained previously for the 3D case. In the 3D case the correction from the non-analytic (*i.e.*, singular) point is found to be small, of the order $\sim \hbar \omega_D / E_F$ with ω_D = Debye frequency. In our case, the major change of $N(E)$ as a function of x may well determine the observed characteristic dramatic variation of T_c with x.

From the BCS theory of superconductivity, the transition temperature T_c is determined by the pairing interaction strength, $\lambda = N(E_F)V$ and V is the pairing potential arising from the electron-phonon interaction. Considering the transition temperature T_c as a function of x, we can write down the dependence of T_c on x as follows

$$\frac{dT_c}{dx} = \frac{\partial T_c}{\partial N(E_F)} \frac{dN(E_F)}{dx} + \frac{\partial T_c}{\partial V} \frac{dV}{dx}. \tag{1}$$

this assumes that the average phonon frequency remains constant. When

$$\left| \frac{1}{N(E_F)} \frac{dN(E_F)}{dx} \right| \gg \left| \frac{1}{V} \frac{dV}{dx} \right|$$

the change ΔT_c will be determined by the change of $N(E_F)$. Since the change of Fermi surface topology due to the SPS has a small (in our terms) effect on the phonon spectrum,[15] the variation of the pairing potential V will be small.[16] (Indeed, the frozen phonon calculations, which yield a high frequency for the breathing mode and a large electron-phonon interaction energy, find little dependence in Fermi surface nesting.[11]) Under these approximate conditions, the change in T_c with composition x is associated with a change of $N(E_F)$. In fact, recent reports[14] show a large variation of T_c *vs.* x which rises sharply from 0 K as $x < 0.03$ to 22 K at $x = 0.08$, hits a maximum at $x \simeq 0.15$ and then drops off sharply. These results are very consistent with our picture. Thus, it is clear that the strong variation in $N(E_F)$, derived from the quasi-2D van Hove singularity (*cf.*, Fig. 1 insert), plays a dominant role in the anomalous behaviour of T_c as a function of x.

One of the traditional devices for obtaining information about the interaction driving the superconductivity is to measure the electronic contribution to the specific heat, γ, and to obtain the dressed DOS at E_F, $N^\gamma(E_F)$. Writing $N^\gamma(E_F) = N(E_F)(1 + \lambda)$ where $N(E_F)$ is the non-interacting (or band) DOS, one can, from the measured $N^\gamma(E_F)$ and calculated $N(E_F)$, obtain the interaction parameter, λ which plays an essential role in superconductivity. Generally, λ

is thought to arise from electron-phonon, λ_{e-ph}, electron-electron, λ_{e-e}, and spin fluctuation contributions, λ_{sp}, (expected to be small in these wide band materials).

Since λ is so important, a number of groups have attempted the measurement of $N^{\gamma}(E_F)$ using a variety of techniques. Unfortunately, the difficulty of the measurement, particularly on the powders available so far, has resulted in a variety of values for $N^{\gamma}(E_F)$. In turn, the use of our calculated $N(E_F)$ values results in a range of derived values for λ. Further, since, as we have seen, $N(E_F)$ is a strong function of divalent composition, x—and of the oxygen vacancy defect concentration— care is required in relating a measured $N^{\gamma}(E_F; x)$ to the corresponding theoretical $N(E_F; x)$ to derive $\lambda(x)$. A further note of caution is to be sounded: our calculated values of $N(E_F; x)$ are obtained using a rigid band approximation to the results obtained in a local density calculation for low temperature lattice parameters measured[17] for a Ba composition, $x = 0.15$. Calculations of $N(E_F)$ for varying measured lattice constants (corresponding to the varying x values) are in progress.

With these caveats, we may cite some of the γ values we have obtained from measured values of γ (in mJ/mole-K^2) known to us: (i) from their dH_{c2}/dT (and resistivity) results (for $x = 0.15$), Kwok, et al.[18] obtain $\gamma_{min} = 4.9$ (dirty limit) and $\gamma_{max} = 7.3$ (clean limit) which gives us $\lambda = 0.07$ and 0.59, respectively. (ii) Battlogg, et al.,[19] cite an average of three different experiments (on $x = 0.15$), $\gamma_{BCS} = 6 \pm 1.5$, which results in $\lambda = 0.3 \pm 0.3$. (iii) From high field measurements, on the dH_{c2}/dT, Foner, et al.,[20] obtain (for $x = 0.15$ and 0.20) $\gamma = 6.275$ (at the resistance mid-point) and 14.8 (at the resistance drop onset temperature); the corresponding λ's have a large range, from 0.39 to 2.2. (iv) The results of Panson, et al.,[21] (for $x = 0.2$) yield a γ value (2.7 ± 0.1) which is far below our bare value. (v) Experiments by Thiel, et al.,[22] at a sample with $x = 0.17$ (just at the SPS) derive from dH_{c2}/dT and a resistivity of $353\ \mu\Omega$-cm, a $\gamma = 6.3$ and hence $\gamma = 0.39$. This range of γ values reflects the difficulties in the measurements cited above. Excluding the upper value of Foner, et al., one sees a consistent set of γ values emerge namely $\lambda = 0.3 \pm 0.3$. This result indicates that one may well be in the weak coupling limit.

As stated above, a recent frozen phonon calculation[11] found that (due to the strong Cu d-O(1) p bounding) the optic breathing mode has a very high frequency ($\theta_D = 1100\ \text{K}$). If one uses the measured T_c ($\sim 40\ \text{K}$) and this calculated θ_D in a simple BCS equation with λ and μ^*, one finds that $\lambda - \mu^* = 0.3$, or well within the range of values for λ given above. We can also estimate theoretically the contribution from the optic breathing mode to λ. Using $\langle u^2 \rangle = \frac{\hbar}{2M\omega_q}$, we obtain an r.m.s. phonon displacement $\simeq 0.04\ \text{Å}$. The electron-phonon interaction matrix element can be calculated from the electronic band splitting near the phonon BZ boundary due to the phonon induced O-displacement; a deformation potential $\simeq 3\ \text{eV/Å}$ is obtained for states with Cu $d_{x^2-y^2}$-O $p_{x,y}$ character near E_F. These quantities, together with the calculated $N(E_F)$, give $\lambda_{\vec{q}}$ (breathing mode) $\simeq 0.25$. Notice that, despite its high frequency, the breathing mode also makes a sizable contribution to the Fermi surface averaged value of λ.

Further, using our calculated v_F values, a crude (order of magnitude) estimate may be made of the upper critical fields, $H_{c2\parallel}$ and $H_{c2\perp}$ and coherence lengths $\xi_{\parallel}$

and $\xi_\perp$. While not appropriate, we use the clean limit[23] since we have no information about the scattering length. We find that $\xi_\parallel \gtrsim 130$ Å, $\xi_\perp \simeq 1$ Å, and $H_{c2\parallel} \simeq 200T$, $H_{c2\perp} \simeq 2T$—clearly indicating the effects of the strongly 2D nature of the calculated band structure.

Finally, single crystal measurements are essential for obtaining more meaningful values of the various parameters discussed in this paper.

Note added: Since submission of this paper, there appeared another paper[24] with a similar band structure to that of Ref. 5.

ACKNOWLEDGEMENT

This work was supported by the National Science Foundation Grant No. DMR85-18607.

We are grateful to J.-H. Xu and T. J. Watson-Yang for helpful discussions and to J. Jorgensen and C. Kimball for informing us of their unpublished work.

REFERENCES

1. J. G. Bednorz and K. A. Müller, *Z. Phys. B*, **64**, 189 (1986).
2. S. Uchida, H. Takagi, K. Kitazawa, and S. Tanaka, *Phys. Lett.*, (submitted); H. Takagi, S. Uchida, K. Kitazawa, and S. Tanaka, *Jap. J. Appl. Phys. Lett.*, submitted.
3. C. W. Chu, P. H. Hor, R. L. Meng, L. Gao, Z. J. Huang, and Y. Q. Wang, *Phys. Rev. Lett.*, **58**, 405 (1987).
4. R. J. Caca, R. B. Van Dover, B. Batlogg, and E. A. Rietman, *Phys. Rev. Lett.*, **58**, 408 (1987).
5. Jaejun Yu, A. J. Freeman, And J.-H. Xu, *Phys. Rev. Lett.*, **58**, 1035 (1987)
6. H. J. F. Jansen and A. J. Freeman, *Phys. Rev. B*, **30**, 561 (1984); E. Wimmer, *et al.*, *Phys. Rev. B*, **24**, 864 (1981).
7. J.-H. Xu, T. J. Watson-Yang, Jaejun Yu, and A. J. Freeman, *Phys. Lett.*, to appear.
8. I. M. Lifshitz, *Soviet Physics JETP*, **11**, 1130 (1960).
9. L. Dagens, *J. Phys. F: Metal Phys.*, **8**, 2093 (1978).
10. As apparently recently observed by C. Kimball, *et al.*, unpublished.
11. C. L. Fu and A. J. Freeman, *Phys. Rev. Lett.*, submitted.
12. M. I. Kaganov, *Sov. Phys. Usp*, **28**, 257 (1985).
13. A. J. Freeman, Jaejun Yu, J. Jorgensen, D. Hinks, and D. Capone, unpublished.

14. S. Kambe, K. Kishio, K. Kitazawa, K. Fueki, H. Takagi, and S. Tanaka, preprint.
15. R. J. Higgins and H. D. Kaehn, *Phys. Rev.*, **182**, 649 (1969).
16. A crude calculation of the electronic contribution to λ, based on the rigid muffin tin approximation for varying x, supports the argument that $\left|\frac{1}{v}\frac{dV}{dx}\right|$ is small all relative to $\left|\frac{1}{N(E_F)}\frac{dN(E_F)}{dx}\right|$, assuming the average phonon frequency is constant.
17. J. D. Jorgensen, *et al.*, *Phys. Rev. Lett.*, to appear.
18. W. K. Kwok, G. W. Crabtree, D. G. Hinks, D. W. Capone, J. D. Jorgensen, and K. Zhang, preprint.
19. B. Batlogg, A. P. Ramirez, R. J. Cava, R. B. van Dover, and E. A. Rietman, preprint.
20. S. Foner, T. P. Orlando, E. J. McHiff, Jr., J. M. Tarascon, L. H. Greene, W. R. McKinnon, and G. W. Hall, *Phys. Rev. Lett.*, submitted.
21. A. J. Panson, G. R. Wagner, A. I. Braginski, J. R. Gavaler, M. E. Janocko, H. C. Pohl, and J. Talvacchio, *Appl. Phys. Lett.*, submitted.
22. J. Thiel, S. Song, K. Poepplemeier, J. B. Ketterson, and A. J. Freeman, to be published.
23. W. E. Lawrence and S. Doniach, in *Proc. 12th Int. Conf. Low Temp. Phys.*, Kyoto, (1970), ed. by E. Kanda, Keigaku, Tokyo, 1971, p. 361.
24. L. F. Mattheiss, *Phys. Rev. Lett.*, **58**, 1028 (1987).

A. J. Freeman, Jaejun Yu, S. Massidda
Department of Physics and Astronomy
Northwestern University
Evanston, Illinois 60208

and

D. D. Koelling
Materials Science Division
Argonne National Laboratory
Argone, Illinois 60349

Electronic Structure, Charge Transfer Excitations, and High Temperature Superconductivity[†]

The high precision local density electronic band structure results (for $YBa_2Cu_3O_7$, $YBa_2Cu_3O_6$, and $GdBa_2Cu_3O_7$) lead to the possibly important role of charge transfer excitations as the mechanism of high T_c superconductivity, and explain the coexistence of magnetism and superconductivity in the high T_c rare-earth superconductors.

INTRODUCTION

The origin of superconductivity (SC) in the new metallic oxides remains a challenge despite some intriguing hints obtained from experiment and electronic structure calculations. Still, it is now quite apparent that understanding the electronic structure and properties of the new high T_c superconductors, $La_{2-x}M_xCuO_4$[1] and $YBa_2Cu_3O_{7-\delta}$[2] is emerging. This is an important step towards achieving an understanding of the origin of their SC. Detailed high resolution local density band

[†]Reprinted from *(Proc. of Yamada Conference)*, *Physica* **148**B, 212–217 1987.

Theories of High Temperature Superconductivity
Addison-Wesley Publishing Company, 1988   **53**

structure results have served to demonstrate what has been our major emphasis, namely the close relation of the physics (band structure) and chemistry (bonds and valences) to the structural arrangements of the constituent atoms[3-4]. They may provide insight into the basic mechanism of their SC. Results obtained on the systems we have studied— La_2CuO_4, $YBa_2Cu_3O_7$, $YBa_2Cu_3O_6$, and $GdBa_2Cu_3O_7$— indicate a number of common chemical and physical features, especially the role of oxygens (and oxygen vacancies), which bear further scrutiny.

$La_{2-x}M_xCuO_4$

In $La_{2-x}M_xCuO_4$, we found early on[5,6] that the highly 2D structure of this layered perovskite was strongly reflected in the electronic structure and properties. For the starting compound La_2CuO_4 in the bct phase, the electronic structure and properties were found to be dominated by the layered in-plane interactions of the $Cud_{x^2-y^2}$ and $O(1)2p$ electrons. The single (essentially 2D) parabolic $d-p$ band crossing E_F gives rise to a single sheet Fermi surface. The results also[5,6] emphasized the remarkable 2D nature of the electronic structure with conductivity confined essentially to the metallic Cu-O(1) planes separated by insulating (ionic) planes of La-O(2) ions. Further, we carried out total energy frozen phonon calculations on $La_{2-x}M_xCuO_4$ to determine the role of the optical breathing mode.[7] Perhaps among our most significant findings were the large charge fluctuations between the in-plane Cu atoms, which can be as large as 0.3 electrons at the maximum of the O displacement. A key role in possible charge transfer excitations (CTE) is played by excitations between occupied localized Cud_{z^2}-$Op_\perp$ and empty itinerant $Cud_{x^2-y^2}$-$Op_\parallel$ states. We emphasized[10] that these could couple resonantly with natural "Cu^{2+}-Cu^{3+}"-like charge fluctuations which exist in the $x > 0$ compounds, with important consequences for the SC.

CHAINS vs. PLANES IN $YBa_2Cu_3O_{7-\delta}$

An intriguing aspect of the origin of the high T_c SC in $YBa_2Cu_3O_{7-\delta}$ is the possible role played by the Cu-O chains vs. the Cu-O planes which make up this structure. While $La_{2-x}M_xCuO_4$ SC consists only of Cu-O planes, there is evidence in $YBa_2Cu_3O_{7-\delta}$ that introducing additional O vacancies into the chains (increasing δ to 0.5 or greater) results in the onset of a tetragonal phase and the lowering of T_c and its eventual absence.[8] Thus, the important structural features of the $YBa_2Cu_3O_{7-\delta}$ compounds are the presence of 2D Cu(2)-O(2,3) planes and 1D Cu(1)-O(1) chains with the O_4 bridging the Cu(1) and Cu(2). These features arise from a total absence of O atoms in the Y-O planes and an ordered absence of O atoms in the Cu-O planes between the Ba-O planes of the perfect triple perovskite

$YCuO_3 \cdot (BaCuO_3)_2$. As a result, there are two $Cu(2)$ ions (Cu^{2+} ions in the simple ionic picture) in five-coordinated positions and one $Cu(1)$ ion (Cu^{3+} ion) in a four coordinated position.[9] Due to the O vacancies in the adjacent Y-O plane, the additional distortions of the $O(2)$ and $O(3)$ ions (the so-called "dimpling") arise—which is chemically taken as a signature of a Jahn-Teller (d^9 Cu^{3+}) ion.

Since the $Cu(1)$-$O(4)$ distance is much shorter than the $Cu(2)$-$O(4)$ distance, the interactions between $Cu(1)$-$O(4)$ are expected to be much stronger than between $Cu(2)$-$O(4)$. In fact, from the calculated DOS, it was demonstrated[7] that the overall features of the DOS contributions from $Cu(2)$, $O(2)$, and $O(3)$ resemble each other as do the DOS from the $Cu(1)$, $O(1)$, and $O(4)$. Further, since the $O(4)$ ions are located in the peculiar position of forming the junction between $Cu(1)$ and $Cu(2)$, they bond to $Cu(2)$ as well as to $Cu(1)$. Now since the bonding of $Cu(1)$-$O(4)$ is stronger than that of $Cu(2)$-$O(4)$, the $O(4)$ contribution to the DOS is similar to that of $Cu(1)$ but still shows vestiges of the interactions with $Cu(2)$.

Our calculated valence band structure[10] of stoichiometric $YBa_2Cu_3O_7$ gives a remarkably simple band structure near E_F arising from a complex set of 36 bands (originating from three Cu_{3d} (and seven O_{2p} atoms). As seen in Fig. 1(a), there are four bands considered to be important. Two strongly dispersed bands crossing E_F consisting of $Cu(2)d_{x^2-y^2}$-$O(2)p_x$-$O(3)p_y$ orbitals in the 2D $Cu(2)$-O planes show 2D character, which is very similar to that of the $Cud_{x^2-y^2}$-$Op_{x,y}$ bands in $La_{2-x}M_xCuO_4$. The symmetry allowed interactions between the $Cu(2)$ bands and the $Cu(1)$ bands result in a complicated dispersion for the $Cu(2)$ bands (as occupied bands) along Γ-X and Γ-Y. Significantly, the $Cu(1)d_{x^2-y^2}$-$O(1)p_y$-$O(4)p_z$ anti-bonding ($dp\sigma$) band shows the (large) 1D dispersion expected from the $Cu(1)$-$O(1)$-$Cu(1)$ linear chains but is almost entirely unoccupied. This 1D conduction band is in sharp contrast to the narrow $dp\pi$ band (formed from the $Cu(1)d_{zy}$-$O(1)p_z$-$O(4)p_z$ orbitals), which is almost entirely occupied in the stoichiometric ($\delta = O$) compound and becomes fully occupied for the superconducting materials ($\delta \geq 0.1$). Since this almost flat π band lies just below and crosses E_F (for $\delta = 0$), it gives rise to peaks in the DOS near E_F making the DOS at E_F, $N(E_F)$, sensitive to the position of E_F (i.e., to δ).

In our calculation for $\delta = 0$, $N(E_F) = 1.13$ states/eV Cu-atom, which is comparable to the 1.2 and 1.9 states/eV Cu-atom for $La_{2-x}Sr_xCuO_4$ at $x = 0$ and at the peak of $x = 0.16$, respectively. For increasing δ values (hence increasing E_F in a rigid band model), $N(E_F)$ decreases sharply to 0.87 states/e!V-Cu atom for $\delta = 0.1$ and to 0.52 states/eV Cu-atom for $\delta = 0.2$ (after which the DOS remains roughly constant). Thus it is clear that the $N(E_F)$ per Cu atom in the high T_c superconductor is lower than in the lower T_c $La_{2-x}M_xCuO_4$ case—contrary to expectations involving the electron-phonon interaction, and in agreement with experiment.

Charge density calculations,[10,11] both for the total valence charge and for the individual states crossing E_F, reflect the structural properties of the material. Charge density plots for the individual states near E_F demonstrate the 2D nature of $Cu(2)$-$O(2)$-$O(3)$ $dp\sigma$ bands and the 1D nature of the $Cu(1)$-$O(1)$-L(4) $dp\sigma$ bands. The ionic Y (or R = rare earth) atoms act as electron donors and do not otherwise participate. Also, the partial DOS at E_F for Y give extremely low values for the

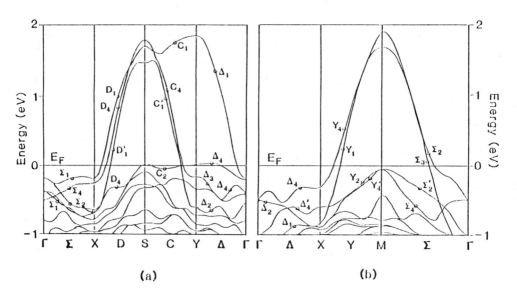

FIGURE 1 Energy bands near E_F of (a) YBa$_2$Cu$_3$O$_7$ and (b) YBa$_2$Cu$_3$O$_6$.

conduction electrons (the same is true for Gd). These results give an immediate explanation for the observed[12] coexistence of the high T_c SC and magnetic ordering in the RBa$_2$Cu$_3$O$_{7-\delta}$ structures. The lack of conduction electron density around the R-site (*cf.* Fig. 2)[13] means that the unpaired rare-earth f-electrons are decoupled from the Cooper pairs (*i.e.*, magnetic isolation) and so cannot pair-break.

YBa$_2$Cu$_3$O7 vs. YBa$_2$Cu$_3$O6

Recently, several neutron experiments[14] showed that the oxygen vacancies concentrate on the O1 sites and change the composition and symmetry from orthorhombic (in YBa$_2$Cu$_3$O$_7$) to tetragonal (in YBa$_2$Cu$_3$O$_6$). The absence of oxygens on the O(1) site destroys the 1D chain structure in the YBa$_2$Cu$_3$O$_6$. The additional oxygen vacancies, therefore, change the local symmetry as well as the electronic configuration around the Cu(1) sites. In this geometry, each Cu(1) ion would be completely isolated from the other Cu(1) ions in the Cu(1) plane (having no oxygens lying between the Cu(1)'s) and remain as Cu$^+$ ions with a completely filled d-shell. Hence, the d-orbital states of Cu(1) are expected to be very localized in the Cu(1) plane. One notable consequence of the change of structure is that the Cu(1)-O(4) distance in YBa$_2$Cu$_3$O$_6$ is even shorter than in YBa$_2$Cu$_3$O$_7$.

To examine these expectations and to provide insight into the possible role[11] of the CTE in the 1D chains of YBa$_2$Cu$_3$O$_{7-\delta}$, we compare the results of calculations[15] for both YBa$_2$Cu$_3$O$_7$ and YBa$_2$Cu$_3$O$_6$, focusing on the role of chains *vs.* planes. As shown in Fig. 1, the calculated band structure (near E_F) of tetragonal YBa$_2$Cu$_3$O$_6$

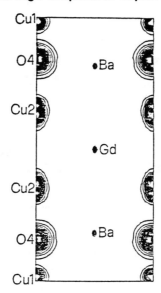

FIGURE 2 Contour plots in the [110] plane of valence charge density near E_F (within $\sim$ 0.2 eV) of GdBa$_2$Cu$_3$O$_7$.

and orthorhombic YBa$_2$Cu$_3$O$_7$ both exhibit two strongly dispersed 2D bands crossing E_F consisting of Cu(2)$d_{x^2-y^2}$-O(2)p_x-O(3)p_y orbitals in the 2D Cu(2)-O planes. However, the dominant 1D electronic structure, arising from the linear chains in YBa$_2$Cu$_3$O$_{7-\delta}$, is completely absent in YBa$_2$Cu$_3$O$_6$. Instead of the 1D structure of $dp\sigma$ and $dp\pi$ states from the Cu(1)-O(1)-O(4) chains, the O$_6$ compound has two bands consisting of Cu(1)$d_{yz}(d_{zx})$-O(4)$p_y(p_x)$ orbitals and are degenerate at the points Γ and M in the BZ.

In addition to the two conduction bands crossing E_F, there is a very small contribution to $N(E_F)$ from the Cu(1)-complex in the linear chains; therefore, the DOS at E_F of YBa$_2$Cu$_3$O$_6$ is reduced to 0.67 states/eV Cu-atom, which is smaller than that of YBa$_2$Cu$_3$O$_7$. In addition to the reduction of $N(E_F)$, there is a strong hybridization of the Cu(2)-O(2(3)) $dp\sigma$ bands and the Cu(1)-O(4) $dp\pi$ bands along the ΓM direction.

In considering magnetic properties, we calculated[11] the Stoner factors for both YBa$_2$Cu$_3$O$_6$ and YBa$_2$Cu$_3$O$_7$. We found a surprisingly large Stoner factor $S = N(E_F)I$, $S = 1.12$, for YBa$_2$Cu$_3$O$_{7-\delta}$ at $\delta = 0$ (indicating a possible magnetic instability), with the largest contribution to S coming from the O(1) and O(4). As δ becomes greater than 0.1, the Fermi level E_F increases and moves above the flat $dp\pi$ band formed mostly by the O(1), O(4), and Cu(1) orbitals, and S drops sharply to 0.86 for $\delta = 0.1$ and 0.52 for $\delta = 0.2$. In contrast, the Stoner factor for YBa$_2$Cu$_3$O$_6$, $S = 1.38$, is even larger. Further, the dominant contribution to S comes from the 2D layers of Cu(2) and O(2(3))k where the 2D conduction bands originate. This suggests a possible magnetic ordering in the 2D Cu(2)-O(2(3)) conducting layer, as a result of localization of the Cu(2)-O(2(3)) hybridized states within each 2D layer due to the isolation of Cu(1) ions, which may relate to its observed semiconductivity.

These results appear to be consistent with the recent susceptibility experiments[20] on both orthorhombic $YBa_2Cu_3O_{7-\delta}$ and tetragonal $YBa_2Cu_3O_6$. In the experiment, $YBa_2Cu_3O_{7-\delta}$ shows Pauli susceptibility, but $YBa_2Cu_3O_6$ shows a Curie-Weiss behavior.

MECHANISM OF SUPERCONDUCTIVITY

Many authors have discussed the excitonic mechanism[17] of SC, in which the effective attractive interaction between conduction electrons originates from virtual excitations of excitons rather than phonons. The basic idea of the models proposed is that conduction electrons residing on the conducting filament (or plane) induce electronic transitions on nearby easily polarizable molecules (or complexes), which result in an effective attractive interaction between conduction electrons. As perhaps a striking realization of the excitonic mechanism of SC, $YBa_2Cu_3O_{7-\delta}$ has two 2D conduction bands and additional highly polarizable 1D electronic structure between the two conduction planes.

We have made crude estimates[11] of the electron-phonon interaction in $YBa_2Cu_3O_{7-\delta}$, using the rigid muffin-tin approximation (RMTA) to calculate the McMillan Hopfield constant η and the electron-phonon coupling constant, λ. Surprisingly, by far the largest contribution to η comes from the O(1) atoms, again indicating the important role played by the "metallized" oxygens, We find that the calculated T_c, even for the assumed Debye temperature $\theta_D = 100\,\mathrm{K}$, cannot exceed the value of $\sim 32\,\mathrm{K}$—a similar limit found previously in $La_{2-x}M_xCuO_4$.[18] Despite the crudeness of the RMTA approach, it is expected that this estimate cannot be so far off (*i.e.*, a factor of three) and so casts doubt on a purely electron-phonon explanation of the observed high T_c.

On the other hand, it is expected that the Cu(1)-O(1)-O(4) chain can play a critical role in the origin of SC. A simple rigid band treatment of the band results[11] suggests that the Cu(1)d_{zy}-O(1)p_z-O(4)p_y ($dp\pi$ anti-bonding) band becomes fully occupied for $\delta \geq 0.1$. Hence, excitations to its (almost) empty Cu(1)$d_{z^2-y^2}$-O(1)p_y-O(4)p_z ($dp\sigma$ anti-bonding) partner band can create strong polarization fields because the $dp\pi$ state is highly localized whereas the $dp\sigma$ state is fairly itinerant along the Cu(1)-O(1) chains. As a result of the difference in the bonding character of the $dp\pi$ and $dp\sigma$ states, a "local" charge transfer excitation (CTE) from the $dp\pi$ state to $dp\sigma$ state may lead to significant electronic polarization. Incorporating the interactions between the 2D conduction electrons and the charge transfer excitations (excitons), could produce the pairing interactions, via an exchange of excitons, which would enhance the T_c.

We previously discussed[11] the importance of the 1D feature in the electronic structure near E_F pointing out the possible role played by charge transfer excitations ("excitons") of occupied (localized Cu(1)-O $dp\pi$ orbitals into their empty (itinerant) Cu(1)-O $dp\sigma$ anti-bonding partners. As shown schematically in Fig. 3, we can characterize the 1D electronic structure with two types of electronic states in it, one free-electron-like (the well-dispersed $dp\sigma$ band) and the other localized (the almost flat $dp\pi$ state). When the localized hole is created (effectively a "Cu^{4+}-complex") due to the excitation, a strong attractive correlation between the hole and excited electron may lead to an electron-hole bound state ("exciton"). Hence, this excitation of the localized $dp\pi$ to the extended $dp\sigma$ with the electron-hole correlation in the 1D electronic structure will give rise to a strong polarization in the 1D chains between two conduction planes and couple to the 2D conduction electrons, which carry most of the SC.

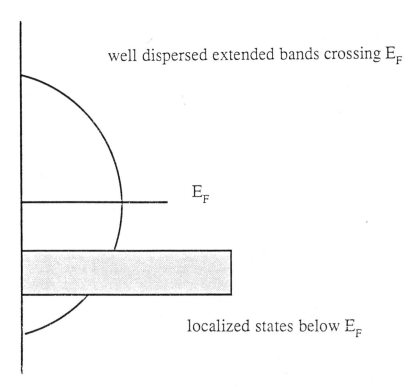

well dispersed extended bands crossing E_F

E_F

localized states below E_F

FIGURE 3 Schematic drawing of the 1D electronic structure of $YBa_2Cu_3O_7$.

Finally we should note that while we discussed the CTE based upon the excitations of occupied *localized* $dp\pi$ states into the empty *itinerant* $dp\sigma$ anti-bonding band, Varma, *et al.*[19] proposed a charge transfer resonance mechanism by considering excitations of the bonding $Cu d_{x^2-y^2}$-$Op_x - p_y$ $dp\sigma$ state (β) and into its anti-bonding (conduction) $dp\sigma$ band (α) partner. Starting from the particle-hole excitation $\beta \rightarrow \alpha$, they argued that a sharp excitonic resonance from the $\alpha\beta$ transitions can happen if the $\alpha\beta$ interband transition gap, which is more than 6 eV. However, it appears unreasonable to neglect the presence of the $dp\pi$ orbital states which lie near E_F, and hence $dp\pi \to dp\sigma$ particle-hole excitation must be included in the CTE model. We again note that similarly in the case of $La_{2-x}Sr_xCuO_4$, the "Cu^{2+}-Cu^{3+}"-like resonant charge fluctuations induced by the optic breathing mode were shown[7] to lead to transitions between the in-plane $d_{x^2-y^2}$ orbitals and localized out-of-plane dz^2 orbital states located just below E_F to the itinerant in-plane $d_{x^2-y^2}$ orbitals. Finally, the special role of O-p electrons involved in the photoemission process (which may relate to the excitonic mechanism) has been discussed in detail for both the $La_{2-x}M_xCuO_4$[20] and $YBa_2Cu_3O_{7-\delta}$[21] cases.

ACKNOWLEDGEMENT

Work supported by the NSF (DMR Grant No. 85-20280) through the NU MRC and DMR Grant No. 85-18607, and a computing grant from its Division for Advanced Scientific Computing, and by the DOE (Contract No. W-31-109-ENG-38). We are grateful to NASA Ames personnel in the NAS program for help with the use of their CRAY2. We regret that space limitations did not permit full referencing to the extensive work in this exciting field.

REFERENCES

1. J. G. Bednorz and K. A. Müller, *Z. Phys. B*, **64**, 189 (1986).
2. M. K. Wu, *et al.*, *Phys. Rev. Lett.*, **58**, 908 (1987).
3. R. J. Cava, *et al.*, *Phys. Rev. Lett.*, **58**, 1676 (1987).
4. D. G. Hinks, *et al.*, *Appl. Phys. Lett.*, **50**, 188 (1987).
5. J. Yu, *et al.*, *Phys. Rev. Lett.*, **58**, 1035 (1987).
6. L. F. Mattheiss, *Phys. Rev. Lett.*, **58**, 1028 (1987).
7. C. L Fu and A. J. Freeman, *Phys. Rev. B*, **35**, 8861 (1987).
8. I. K. Schuller, *et al.*, *Solid State Commun.*, **63**, 385 (1987).
9. M. A. Beno, *et al.*, *Appl. Phys. Lett.*, **51**, 57 (1987).
10. S. Massidda, *et al.*, *Phys. Lett.*, **122**, 198 (1987).
11. J. Yu, *et al.*, *Phys. Lett.*, **122**, 203 (1987).
12. J. O. Willis, *et al.*, *J. Magn. Matls.*, **67**, L139 (1987).
13. J. Yu and A. J. Freeman, to be published.
14. A. Santoro, *et al.*, *Mat. Res. Bull.*, **22**, 1007 (1987); J. D. Jorgensen, *et al.*, *Phys. Rev. B*, in press.
15. J. Yu and A. J. Freeman, to be published.
16. M Ishikawa, *et al.*, preprint.
17. W. A. Little, *Phys. Rev.*, **134**, A1416 (1964); V. L. Ginzburg, *JETP*, **46**, 397 (1964).

18. J. Yu and A. J. Freeman, unpublished; W. E. Pickett, *et al., Phys. Rev. B,* **35** 7252 (1987).
19. C. M. Varma, *et al., Solid State Commun.,* **62**, 681 (1987).
20. J. Redinger, *et al., Phys. Lett.,* in press.
21. J. Redinger, *et al., Phys. Lett.,* in press.

Guang-Lin Zhao, Yongnian Xu, W. Y. Ching
Department of Physics
University of Missouri-Kansas City
Kansas City, Missouri 64110

and

K. W. Wong
Department of Physics and Astronomy
University of Kansas
Lawrence, Kansas 66405

Theoretical Calculation of Optical Properties of Y-Ba-Cu-O Superconductors[†]

The optical properties of orthorhombic $YBa_2Cu_3O_7$ crystal are studied by a first-principles method. The interband optical conductivity shows strong directional anisotropy in the 0–3.0 eV range. A plasmon energy of 2.8 eV is predicted. Analysis of the static dielectric constant in the case of a semiconductorlike band structure supports the excitonic-enhanced superconducting mechanism in the Y-Ba-Cu-O system.

The discovery of the superconducting La-M-Cu-O and Y-Ba-Cu-O systems with superconducting transition temperatures T_c above 35 and 90 K, respectively, has spurred a great deal of experimental and theoretical investigation on these materials.[1-5] The nature of electron states in these materials plays a key role in understanding the properties and the possible superconducting mechanism in these complex ceramic oxides. The electronic structure of $YBa_2Cu_3O_7$ has been studied by the linearized augmented plane wave method[6-8] and the linear combination of atomic orbitals (LCAO) method.[9] In this paper, we report the first calculation on the optical properties of $YBa_2Cu_3O_7$ crystal in order to gain a deeper understanding of the unusual electronic structure of this material and to provide

[†]Reprinted from *Phys. Rev. B*, **36**, 7203 (1987).

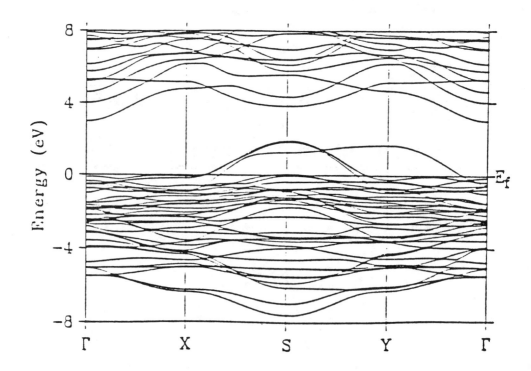

FIGURE 1 Energy bands of YBa$_2$Bu$_3$O$_7$

further evidence of the possible superconducting mechanism in YBa$_2$Cu$_3$O$_{7-\delta}$. Because of the difficulty in obtaining sufficiently large single-phase crystals and of performing the proper surface treatment, no experimentally measured optical data on YBa$_2$Cu$_3$O$_{7-\delta}$ in the ultraviolet or visible frequency range exists in the literature. Optical measurements in the far-infrared region[10] and Raman spectra[11] on YBa$_2$Cu$_3$O$_{7-\delta}$ have been reported. Only very recently additional results on optical measurements on the Y-Ba-Cu-O system in a more extended frequency range have appeared.[12–15]

Our theoretical calculation of optical properties is done for YBa$_2$Cu$_3$O$_7$ in the orthorhombic phase using the crystal-structure data of Beno, *et al.*[16] The band structure of this crystal has been calculated using the first-principles self-consistent orthogonalized LCAO method and detailed results will be reported elsewhere in a longer paper.[17] The band structure along the symmetry lines Γ-X-S-Y-Γ of the lower plane of the Brillouin zone (BZ) is shown in Fig. 1. The band structure in the upper plane (Z-U-R-T-Z) is almost identical and there is very little band dispersion along the vertical k_z direction. Using the wave functions obtained at 147 **k** points in $\frac{1}{8}$ of the BZ, the momentum matrix elements $\langle \psi_n(k)|\mathbf{P}|\psi_l(\mathbf{k})\rangle$ between the occupied state $\psi_n(\mathbf{k})$ and the unoccupied state $\psi_l(\mathbf{k})$ for $|E_n(\mathbf{k}) - E_l(\mathbf{k})| < 10.0$ eV were

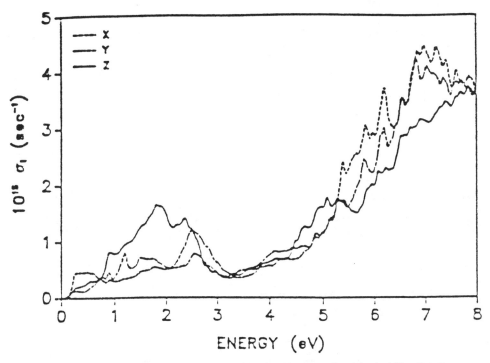

FIGURE 2 x, y, z components of interband optical conductivity in $YBa_2Cu_3O_7$.

evaluated. The interband optical conductivity was then calculated in the dipole approximation according to the Kubo-Greenwood formalism:[18]

$$\sigma_l(\omega) = \frac{2\pi e^2 \hbar}{3m^2 \omega \Omega} \sum_{n,l} \int d\mathbf{k} |\langle \psi_n(\mathbf{k})|\mathbf{P}|\psi_l(\mathbf{k})\rangle|^2 f_l(\mathbf{k})$$

$$[1 - f_n(\mathbf{k})] \delta(E_n(\mathbf{k}) - E_l(\mathbf{k}) - \hbar\omega). \tag{1}$$

The linear analytic tetrahedron method was used for the BZ integration in (1) based in 432 tetrahedrons generated by the 147 $\mathbf{k}$ points in the irreducible portion of the BZ. No attempt was made to include the effect of finite-lifetime broadening. Because the crystal is highly anisotropic, it is necessary to present the σ_{xx}, σ_{yy}, σ_{zz} components of σ_I, which are shown in Fig. 2. It is evident that σ_{zz} is very much different from σ_{xx} and σ_{yy} in the region of 0–3 eV which is related to the quasi-two-dimensional; characteristic of the electronic structure of $YBa_2Cu_3O_7$. The absorption in the z direction is much larger than in either the x or y direction in this region. A crossover occurs at about 5.2 eV, above which the absorption in the z direction becomes smaller than in the perpendicular directions. There are some subtle differences between σ_{xx} and σ_{yy}, as well as in this energy range, reflecting the presence of the one-dimensional O_1-Cu_1 chain along the $\mathbf{b}$ axis and the absence of O atoms along the $\mathbf{a}$ axis in the $z = 0$ plane. It is difficult to associate the structures in the optical conductivity curve with transitions involving specific

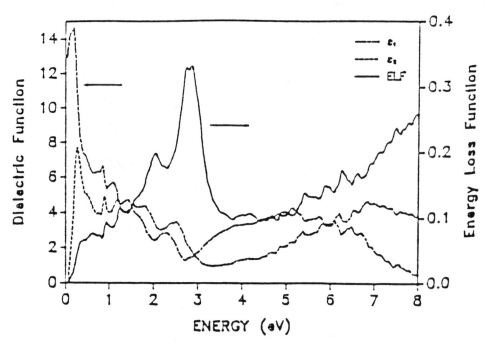

FIGURE 3 x, y, z components of the square of the interband optical matrix elements in YBa$_2$Cu$_3$O$_7$.

pairs of bands because of the multiplicity in the manifold band structures. In order to investigate further the nature of optical anisotropy in YBa$_2$Cu$_3$O$_7$, we plot in Fig. 3 the energy dependence of the square of the components of the optical matrix elements in the same energy range. Below 5 eV, the variation of the squares of the components is much smaller than the variation in the optical-absorption spectra of Fig. 2. Thus, the anisotropy in the optical-absorption is not entirely due to the symmetry of the wave functions of the states involved in the optical transitions, as reflected by the selection rules, but is mostly due to the distribution of the states of different symmetries near the Fermi level and in the vicinity of the semiconductor-like gap. The peak at 2 eV in Fig. 2 is a direct reflection of the transitions from the high-density-of-states (DOS) region between 0 to −2.0 eV to the relatively flat, low-density hole region form 0–1.9 eV.

From the interband optical conductivity, the real and imaginary part of the dielectric function, ϵ_1 and ϵ_2 were evaluated through the Kramers-Kronig relation. From ϵ_1 and ϵ_2, the energy-loss function (ELF) $F(\omega)$ defined as

$$F(\omega) = -\mathrm{Im}\left(\frac{1}{\epsilon}\right) = \frac{\epsilon_2(\omega)}{\epsilon_1^2(\omega) + \epsilon_2^2(\omega)}, \qquad (2)$$

was extracted (see Fig. 4). $F(\omega)$ has a well-defined peak at $\omega_p = 2.8$ eV which is the plasmon frequency for YBa$_2$Cu$_3$O$_7$. This plasma frequency is much smaller than those of a typical semiconductor such as Si ($\omega_p \approx 17$ eV). This plasmon peak has a full width at half maximum of 0.9 eV and there is another smaller peak at 2.0 eV. From Fig. 3, it is clear that this value of the plasmon frequency is due to the fact that ϵ_1 and ϵ_2 have minima at roughly 2.7 and 3.3 eV, respectively. The reason for the minimum in ϵ_2 can be traced to the semiconductorlike band structure of YBa$_2$Cu$_3$O$_7$ shown in Fig. 1. Although the Fermi surface is located about 1.9 eV below the top of the valence band at S, there exists a well-defined gap above S. The bottom of the conduction band is at Γ with an indirect band gap of 1.06 eV and a direct band gap of 1.54 eV at S. The unoccupied region below the top of the valence band (VB) accommodates exactly four holes, and transition of occupied electrons to this hole region accounts for most of the low-energy optical absorption below 3 eV. The presence of the gap between 1.9–3.0 eV above E_f results in a minimum in the optical absorption near 3 eV. and ultimately given rise to a plasmon peak at 2.8 eV. The existence of another plasmon peak at higher energy, where ϵ_1 cuts through zero, should not be ruled out. A recent muon-spin-relaxation measurement[19] indicates that ω_p in YBa$_2$Cu$_3$O$_{7-\delta}$ is probably in the range of 1.2–1.6 eV. In analyzing the optical transition data in the high-frequency range, Sulewski, *et al.*,[14] found they have to use the plasmon frequency value of 2.6 eV to obtain a good fit. Similarly, Orenstein *et al.*,[12] used $\omega_p = 3.0$ eV and obtained an excellent fit to their reflectivity spectrum up to 3 eV. These ω_p values are fully consistent with our calculated value of 2.8 eV.

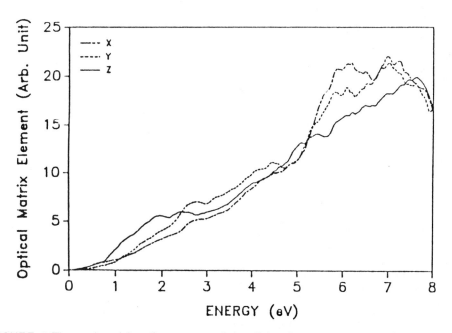

FIGURE 4 The real and imaginary parts of the dielectric constants ϵ_1 and ϵ_2 and the electron energy-loss function in YBa$_2$Cu$_3$O$_7$.

The intraband optical transition is more difficult to treat. Far-infrared optical data from different workers[12–15] are not all consistent and have led to different interpretations. A lot of these inconsistencies have to do with different sample conditions and different ways to correct them. Kamaras, *et al.*,[15] concluded that the frequency-dependent conductivity is decidedly non-Drude in form while the reflectivity spectrum of Wrobel, *et. al.*,[13] shows a smooth, Drude-like decrease over the frequency range up to 700 cm^{-1}. But Orenstein, *et al.*,[12] are of the opinion that optical transitions in both $La_{2-x}Sr_xCuO_4$ and $YBa_2Cu_3O_{7-\delta}$ cannot be described by the nearly free-electron model. Sulewski, *et al.*,[14] argued that free carriers are present and conductivity can be analyzed in terms of the Drude model with a frequency-dependent scattering rate. All these indicate that electron states of the carriers near the Fermi energy in $YBa_2Cu_3O_{7-\delta}$ are very different from those of a conventional free-electron-like metal such as copper. Obviously, more theoretical work is needed to address the intraband optical process in this class of materials where the carriers may be both electrons and holes with large effective masses. Nevertheless, even though the intraband transitions from both electron-electron and hole-hole contribution may be present, it is probably limited to a very low-energy range (< 0.1 eV). The interband transitions should dominate the whole frequency range above the far-infrared region.

The static dielectric function ϵ_0 of $YBa_2Cu_3O_7$ can be obtained from the sum rule

$$\epsilon_0 = \epsilon_1(0) = 1 + \frac{2}{\pi} \int_0^\infty \frac{\epsilon_2(\omega')}{\omega} \, d\omega'. \tag{3}$$

For pure interband transition averaged over the three directions, a value of 12.9 for ϵ_0 is obtained. This value is lower than most of the semimetals. If a finite value of intraband contribution at $\omega = 0$ is to be included, the value of ϵ_0 will be increased. In evaluating Eq. (3), the upper integral limit is set at 15 eV, however, because of the ω^{-1} dependence of the integrand and the fact that the oscillator strength starts to drop for energy higher than 8 eV, it is estimated that the higher-frequency contribution of the integral in Eq. (3) is generally less than 1.5%.

In a recent Letter,[9] we presented an argument for an excitonic-enhanced high-T_c superconducting mechanism in $YBa_2Cu_3O_{7-\delta}$. The original model was suggested more than two decades ago by Jerome, Rice, and Kohn[20] and was subsequently studied in considerable detail by Wong and co-workers.[21,22] The idea is based on the formation of excitons between electrons in the conduction-band (CB) minimum and holes at the maximum of the unfilled VB such that the binding energy $|E_B|$ is slightly larger than E_g, the "band gap" in the semiconductor-like band structure. The excitation gap Δ_{ex}, due to the formation of exciton, enhances the BCS gap Δ_{BCS} to obtain the total excitation gap according to $\Delta_T^2 = \Delta_{BCS}^2 + \Delta_{ex}^2$, thus providing a possible mechanism for high T_c. Since $|E_B|$ is approximately given by [23]

$$|E_B| \approx hc(\mu^*/m) \left(\frac{1}{\epsilon_0^2}\right) \mathbf{R}, \tag{4}$$

where $\mathbf{R}$ is the Rydberg constant, and μ^* is the effective mass of the exciton, μ^* can be accurately estimated from the calculated band structure and E_B must be very close to E_g, so it was concluded that ϵ_0 for $YBa_2Cu_3O_{7-\delta}$ should be in the range of 3–7 in order for the excitonic-enhancement mechanism ot be operational. It should be pointed out that ϵ_0 in (4) is the pure interband part associated with a semiconductorlike band structure with a well-defined "gap." The presence of O defect in $YBa_2Cu_3O_{7-\delta}$, or the replacement of O by F,[24] makes the materials superconducting with high T_c. In the absence of detailed electronic structure and optical calculation for $YBa_2Cu_3O_{7-\delta}$ or Y-Ba-Cu-F-O system, it is fair to conclude that the effect of these on on the band structure of $YBa_2Cu_3O_7$ is to raise the Fermi level, thereby reducing the hole region near the top of the unfilled VB and making the band structure more semiconductor-like. In order to estimate the possible value of ϵ_0 for such systems, we have repeated our calculation of $\epsilon_2(\omega)$ (and hence ϵ_0) by shifting E_F to be at the top of VB at S, while keeping all the states unchanged. We have obtained a value of 3.21 for ϵ_0. Thus, it appears that the interband part of the static dielectric function of the high-T_c oxide systems falls into the range required for the excitonic-enhanced superconducting mechanism to be operational.

In conclusion, we have calculated the interband optical properties of $YBa_2Cu_3O_7$ from first principles. The results show a strong anisotropic effect due to the two-dimensional characteristic of the band structure. From the computed electron-energy-loss spectra, a plasmon frequency of 2.8 eV is predicted. Analysis of the interband part of the static dielectric function for a semiconductor-like band structure supports the excitonic-enhanced superconducting mechanism for $YBa_2Cu_3O_{7-\delta}$.

ACKNOWLEDGEMENTS

We thank Professor M. R. Querry for a number of fruitful discussions. This work is supported by the U. S. Department of Energy Grant No. DE-FG02-84ER45170.

REFERENCES

1. J. G. Bednorz and K. A. Müller, *Z. Phys. B*, **64**, 189 (1986).
2. C. W. Chu, *et al.*, *Phys. Rev. Lett.*, **58**, 405 (1987).
3. R. J. Cava, *et al.*, *Phys. Rev. Lett.*, **58**, 408 (1987).
4. M. K. Wu, *et al.*, *Phys. Rev. Lett.*, **58**, 908 (1987).
5. P. H. Hor, *et al.*, *Phys. Rev. Lett.*, **58**, 911 (1987).
6. S. Massida, J. Yu, A. J. Freeman, and D. D. Koelling, *Phys. Lett. A*, **122**, 198 (1987).

7. Jaejun Yu, S. Massida, A. J. Freeman, and D. D. Koelling, *Phys. Lett. A*, **122**, 203 (1987).

8. L. F. Mattheiss and D. R. Hamann, *Solid State Commun.*, **63**, 395 (1987).

9. W. Y. Ching, Y. Xu, G.-L. Zhao, K. W. Wong, and F. Zandiehnadem, *Phys. Rev. Lett.*, **59**, 1333 (1987).

10. D. A. Bonn, *et al.*, *Phys. Rev. Lett.*, **58**, 2249 (1987).

11. B. Batlogg, *et al.*, *Phys. Rev. Lett.*, **58**, 2333 (1987).

12. J. Orenstein, *et al.*, *Phys. Rev. B*, **36**, 729 (1987).

13. J. M. Wrobel, *et al.*, *Phys. Rev. B*, **36**, 2368 (1987).

14. P. E. Sulewski, *et al.*, *Phys. Rev. B*, **36**, 2357 (1987).

15. K. Kamaras, *et al.*, *Phys. Rev. Lett.*, **59**, 919 (1987).

16. M. A. Beno, *et al.*, *Appl. Phy. Lett.*, **51**, 57 (1987).

17. Y. Xu, G.-L. Zhao, F. Zandiehnadem, and W. Y. Ching, unpublished.

18. D. A. Greenwood, *Proc. R. Phys. Soc. London Ser. A*, **71**, 585 (1958).

19. D. R. Harshman, *et al.*, *Phys. Rev. B*, **36**, 2386 (1987).

20. D. Jerome, T. M. Rice, and W. Kohn, *Phys. Rev.*, **158**, 462 (1967).

21. K. W. Wong and K. K. Bajaj, *Phys. Lett.*, **26A**, 54 (1968); K. W. Wong and S. C. Lo, *ibid.*, **31A**, 260 (1970).

22. S. C. Lo and K. W. Wong, *Nuovo Cimento*, **10B**, 361 (1972); **10B**, 383 (1972).

23. A. N. Kozlov and L. A. Maksimov, *Zh. Eksp. Teor. Fiz.*, **48**, 1184 (1965) [*Sov. Phys. JETP*, **21**, 790 (1965)].

24. S. R. Ovshinsky, *et al.*, *Phys. Rev. Lett.*, **28**, 2579 (1987).

John P. Ralston
Department of Physics and Astronomy
University of Kansas
Lawrence, Kansas 66045

Model Independent Tests of Excitonic Enhancement in High-T_c Superconductors†

We point out simple and model independent tests of proposals that excitonic condensate formation is important for high-T_c superconductors. Elementary calculations predict the features of sharp infrared spectra, and we discuss novel features of the superconducting phase transition. We present evidence for excitonic level transitions in the superconductor $YBa_2Cu_3O_{7-y}$, showing a Rydberg spectrum fit to sharp infrared structure in the $y = 0.1$ superconducting composition in the normal state that is absent in the $y = 0.8$ non-superconducting phase.

The unexpected experimental discoveries of high critical temperature superconductivity[1-3] have stimulated theoretical interest in novel mechanisms for the phenomena. Inasmuch as the discoveries were not anticipated by conventional modeling, a phenomenological approach to modestly probe the important degrees of freedom in a model-independent way seems to be called for. Among the proposed mechanisms, the enhancement of collective effects with a condensed excitonic phase was discussed some time ago[4-6] and has received recent new attention.[7-8] Here we will point out some direct experimental consequences of such excitonic enhancement

†Reprinted from *Phys. Rev. B*, **36**, 8783 (1987).

(ee), which would be of sufficiently dramatic character to immediately settle the importance of ee if observed.

We take the approach of separating the issues of participation of excitons in the superconductivity from detailed modeling of the superconductivity itself. The latter is a deeply non-perturbative problem which can be adequately motivated but probably not established by current theoretical ideas. The calculations of Ching, et al.,[7] provide an example of a detailed motivation for ee in the Y-Ba-Cu-O systems, as do similar discussions of Massida, et al.,[8] Both these and other studies imply the existence of an excitonic condensate in the high-T_c materials; it is this fact we propose to test. We outline three tests for the ee mechanism as follows:

INFRARED ABSORPTION SPECTRA

The excitonic bound states[9] should have discrete energy levels E_n given roughly by

$$E_n \sim -hc \left(\frac{\mu^*}{m}\right) \frac{1}{\epsilon_0^2} \frac{R^y}{n^2} \tag{1}$$

where μ^* is the reduced effective mass of the electron-hole pair, R_y is the Rydberg constant and ϵ_0 is an effective (frequency dependent) dielectric constant. If an exciton binding energy E_B is large enough, there is a tendency for excitons to condense, given other non-perturbative collective requirements.[4-7] To participate in ee with sufficiently de-localized excitons. Ching, et al. argue[7] that $|E_B| \gtrsim E_g$ is needed, with E_g a gap energy of order 1 eV. The condition $|E_B| \gg E_g$ is not acceptable in general because the excitons would be too tightly bound to enhance the effective coupling into Cooper pairs. Inasmuch as E_B parameterizes a collective quantity associated with the long-range order of condensate formation, it is reasonable to use the static dielectric constant $\epsilon_0(0)$ to estimate $E_B \cong E_1(\epsilon_0(0))$. In that case one needs $\epsilon_0(0) \cong 2 - 5$, based on a band gap of order 1 eV.

However the condensed excitons may be excited internally, probing the dynamics of the electron-hole localized states. If (1) applies then one should see sharp infrared structure, with lines at wave numbers k_{ij} given by

$$k_{ij} = R_y^e \left[\frac{1}{n_i^2} - \frac{1}{n_j^2}\right]; \quad R_y^e = \frac{\mu^*}{m} \frac{R_y}{\epsilon_0^2(\overline{k})}, \tag{2}$$

where R_y^e is an effective excitonic Rydberg constant. In (2) we emphasize the $1/n^2$ dependence over the R_y^e energy scale, which may be treated as a phenomenological parameter. Naively, the frequency dependence of $\epsilon_0(k)$ does not affect (2), which assumes as a first approximation that an effective value $\overline{k}$ can be used in any case.

If absorption lines consistent with the $1/n^2$ dependence are observed, it is clear and obvious evidence of excitons. If absorption consistent with (2) is not observed, however, then the *ee* mechanism could still be operating but becomes much more difficult to test. In the transition from the superconducting to the normal phase, one expects a sudden change in the excitonic population, and a sudden change in the line absorption spectra.

The sudden change in the spectrum across the transition should help eliminate problems due to broadening.[9] The continuum exciton condensate we discuss here does have an intrinsic broadening: if $|\beta_x\rangle$ denotes the state of an exciton localized at a coordinate x, then the condensate states of interest are at least as complicated as momentum states $|\beta_k\rangle$:

$$|\beta_k\rangle = \int \frac{d^3x}{\left(\sqrt{2\pi}\right)^3} e^{i\vec{k}\cdot\vec{x}} |\beta_x\rangle \tag{3}$$

However, the spatial size of the exciton state can dominate in determining the broadening due to the superposition (3). To see this, note that the interaction with the photon absorption Hamiltonian H_γ depends on the terms $\langle\beta_{x'}|H_\gamma|\beta_x\rangle$ which is generally large only for $|x - x'|$ of order the exciton size. Thus, without claiming to have a detailed proof, one anticipates that line broadening may be no more of a problem than in ordinary observations on conventional excitonic states.[9]

By now [10] several experimental groups have reported infrared reflectance spectra in the $100-10,000 \text{ cm}^{-1}$ region of interest. The data of the Florida-McMaster group[11] taken on $YBa_2Cu_3O_{7-y}$ shows several distinct peaks that are present only in the superconducting phase ($y = 0.1$, $T = 105 \text{ K}$) and several peaks that are present in both this and the non-superconducting ($y = 0.8$, $T = 300 \text{ K}$) phase (Fig. 1). While the $y = 0.1$ phase is a superconducting composition, the data is reported for a temperature above the phase transition, in the normal phase. Nevertheless the data will be studied here, although the complete temperature dependence through T_c would be more useful. To analyze this data, a few peaks from the $y = 0.1$ sample at frequencies of 191, 279 and 548 cm^{-1} are assumed to be due to excitonic level transitions. These peaks occur above valleys in the non-superconducting specimen. The peaks show different changes as the critical temperature is crossed.[12] To compare these date to the spectrum (2) without knowing R_y^e or prejudicing the assignment of levels, a graphical method was used (Fig. 2) The graphical fit yields the excitonic Rydberg constant,

$$R_y^e = 3923[1 + 7 \times 10^{-6}(k/\text{cm}^{-1})]\text{cm}^{-1}$$

for $100 \lesssim k \lesssim 10,000 \text{ cm}^{-1}$, where k is the inverse wavelength of the reflected photon. The second term in brackets is a small correction one can surmise may be due to the frequency dependence of the dielectric constant: a change of about 3% in $\epsilon_0(k)$ over the interval $k = 100 \text{ cm}^{-1} - 100,000 \text{ cm}^{-1}$ is consistent with (4). If we estimate $\mu^*/m \simeq \frac{1}{2}$, then (2) and (4) imply an empirical value $\epsilon_0(\bar{k}) \simeq 3.7$, a value compatible with the gap energy from calculations.[13] The fit then predicts several lines which are shown as vertical lines in Fig. (1). This simple procedure provides clear evidence for excitons associated with the superconductivity.

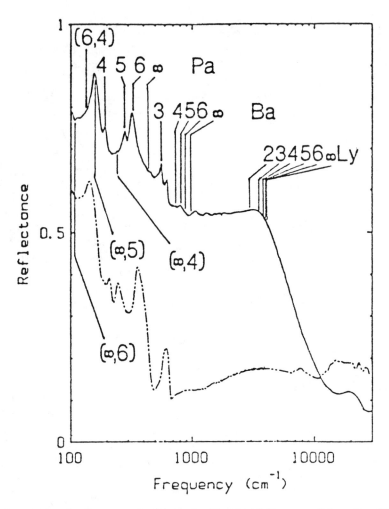

FIGURE 1 Infrared reflectance data from the Florida-McMaster collaboration (Ref. 11) for $y = 0.1$, superconducting, normal phase, (top curve) and $y = 0.8$, non-superconducting (bottom curve) samples. The vertical lines calculated using Eqs. (2, 4) are labeled by initial state integers (n_j) (shown) and final state numbers 1, 2, 3, corresponding to Lyman, Balmer and Paschen series. States of other series are labeled (n_j, n_i).

SUPERCURRENT INFRARED QUENCHING

The ground state to excited state transitions implied by (2) will have the physical effect of promoting excitons out of a condensed state to localized individual states. In a thin film of superconducting material, it may be possible to destroy the super-conductivity by specific resonance-tuned infrared radiation.[14] This is reminiscent

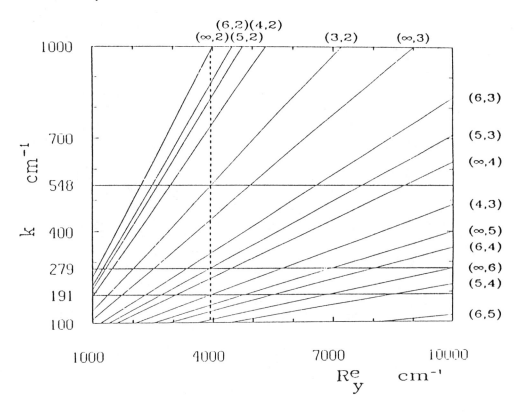

FIGURE 2 Graphical method for assigning levels. Three peaks in the spectrum apparently associated with the superconducting phase only are shown as horizontal lines. Transition energies k_{ij} from Eq. (2), levels (n_i, n_j), are plotted as functions of the parameter R_y^e. The dashed, nearly vertical line fixes R_y^e, Eq. (4).

of the standard procedure of driving Type 1 superconductors into the normal state below T_c by applying a large magnetic field.

For quenching to occur the absorption rate must be larger than the downward transition rate Γ_e from the width of the lines. Thus, for photon-exciton resonance cross section $\sigma_{\gamma e}$ and photon number flux f_γ we need $f_\gamma \sigma_{\gamma e} > \Gamma_e$. To deplete the population throughout the sample of thickness ℓ, the absorption length $L_\gamma = 1/\sigma_{\gamma e}\rho_e$ should satisfy $L_\gamma \gtrsim \ell$, where ρ_e is the exciton number density. A simple relation independent of $\sigma_{\gamma e}$ comes from these conditions, letting the line width $\Delta w \approx \Gamma_e$:

$$\ell \lesssim \ell_{\max}; \quad \ell_{\max} \approx f_\gamma/\Delta w\rho_e.$$

The quenching and dependence of $\ell_{\max}$ on $1/\Delta w$, if observed, would be a clear signal of exciton population depletion.

SPONTANEOUS FLUORESCENCE AT THE PHASE TRANSITION

The formation of an excitonic condensate implies a rather unusual latent heat pathway as excitons are formed. In a sharp transition from above T_c, continuum electrons falling into exciton bound states can release copious amounts of infrared radiation. This is a first order effect in what has traditionally been a second-order system.

The broad character of the transitions in current high-T_c samples may make this difficult to observe. However, a rapidly cooled sample might give rise to a continuum burst of characteristic infrared radiation. The details of the spectrum clearly depend not only on the excitons but also on the band structure of the material.

Taken together, observation of the consequences of *ee* predicted here would be clean-cut and unmistakable. We have emphasized the most elementary and qualitative features which at the same time give most dramatic signals. If the phenomena are consistently observed, the identification of an excitonic condensate as the important degree of freedom in these impressive materials may lead to theoretical control of the high-T_c revolution.

Note added: Temperature dependence of the lines between 100–$500\,\text{cm}^{-1}$ is reported by L. Genzel, *et al.*, (Max-Planck Institut Reports, May 1987). The peaks generally sharpen and increase slightly as the temperature is lowered through T_c.

ACKNOWLEDGMENTS

I thank K. W. Wong for useful discussions and S. N. Sun for help with numerical work. This work was supported in part under Department of Energy Grant No. 85ER40214.A002 and University of Kansas General Research Fund No. 3616–0038.

REFERENCES

1. J. G Bednorz and K. A. Müller, *Z. Phys. B*, **64**, 189 (1986); C. W. Chu, *et al.*, *Phys. Rev. Lett.*, **58**, 405 (1987).
2. M. K. Wu, *et al.*, *Phys. Rev. Lett.*, **58**, 908 (1987).
3. See the listing "High-T_c Superconductivity" in the 25 May issue of *Physical Review Letters*, **58**.
4. D. Jermoe, T. M. Rice and W. Kohn, *Phys. Rev.*, **158** 462 (1967).
5. K. W. Wong and K. K. Bajaj, *Phys. Lett.*, **20**, 651 (1968); K. W. Wong and S. C. Lo, *ibid.*, **31**A, 260 (1970).

6. S. C. Lo and K. W. Wong, *Nu. Cim.*, **10**B, 361, 383 (1972).

7. W. Y. Ching, Y. Xu, G.-L. Zhao, K. Wong, and F. Zandiehnadem, *Phys. Rev. Letters*, **59**, 1333 (1987).

8. S. Massida, J. Yu, A. J. Freeman, and D. D. Koeling, *Phys. Lett.* (1987), submitted.

9. For a review on excitons see R. S. Knox in *Coherent Excitations in Solids*, ed. by B. Dibartolo, Plenum, New York, 1981, p. 183ff.

10. The infrared excitonic peaks were a prediction of the first version of this paper, received before the published receipt date of the data of Ref. 11. The term "excitonic" appears to be used in a generic sense of electronic excitation in that reference, without necessarily implying the particle-hole bound state of our interpretation here.

11. K. Kamaris, *et al.*, *Phys. Rev. Lett.*, **59**, 919 (1987).

12. D. A. Bonn, *et al.*, *Phys. Rev. Lett.*, **58**, 2249 (1987).

13. See, *e.g.*, W. Y. Ching, *et al.*, Ref. 7, and G. L. Zhao, Y. Xu, W. Y. Ching, and K. W. Wong, Kansas preprint (Sept. 1987).

14. It is conceivable that one could achieve supercurrent *enhancement* by the radiation if excitons are involved in an unexpected way. In either case experiments must be done.

15. L. Genzel, *et al.*, Max-Plank Institut Reports, May 1987.

Daniel C. Mattis
Department of Physics
University of Utah
Salt Lake City, Utah 84112

and

Michael P. Mattis
Enrico Fermi Institute
University of Chicago
Chicago, Illinois 60637

Bond Asymmetry and High-T_c Superconductivity

We propose a simple mechanism, anchored in weak-coupling BCS, which ties together the following facts: high T_c; quasi-two-dimensionality; orthorhombic distortion and/or ordered lines of oxygen; proximity to a metal-insulator transition; anomalously small isotope effects. Experimental evidence for the mechanism is discussed. It appears that there is not sufficient evidence to discard the electron-phonon mechanism at the present time.

I. INTRODUCTION

Any list of unusual features of high-T_c superconductors would have to include the following:
 (i) quasi-two-dimensional electronic connectivity;[1,2]
 (ii) the importance of orthorhombic lattice distortion, coupled, in YBaCuO, with ordered lines of oxygen (*the symmetric phase of YBaCuO and LaSrCuO shows no evidence of superconductivity*);[3]
 (iii) Proximity to a metal-insulator transition:
 (iv) anomalously small isotope effects;[4]
 (v) high T_c itself.

Here we show that these features can all be understood in a conventional weak-coupling BCS[5] framework—provided that an important, physically motivated *bond asymmetry term* is included in the Hamiltonian, and that the chemical potential μ is properly tuned. Our mechanism is independent of the specific source of 2-body attraction, and is compatible with a variety of tight-binding models, including the one suggested by band structure calculations.[1,2]

The picture we start from is that of a 2D lattice (*e.g.*, the top or bottom planes of the "triplet" of CuO_2 planes in YBaCuO), with copper on the vertices and oxygen on both the horizontal ('x') and vertical ('y') links, as in Fig. 1. The mechanism is illustrated most clearly in the "toy" model in which the charged carriers (holes[6] in YBaCuO and LaSrCuO) are exclusively in the oxygen bands of the basal planes, *i.e.*, on the links. (An alternative tight-binding model[1,2] including the Cu atoms is discussed below.) We will allow only nearest neighbor motion in our Hamiltonian, so that a carrier on an 'x' link can hop to one of four 'y' links surrounding it, and *vice versa*. In momentum space this becomes

$$H_{\text{hop}}^{xy} = -4t \sum_{\mathbf{k}} \sum_{\sigma=\uparrow,\downarrow} \cos\frac{k_x}{2} \cos\frac{k_y}{2} \left(c_{x\sigma}^{\dagger}(\mathbf{k})c_{y\sigma}(\mathbf{k}) + c_{y\sigma}^{\dagger}(\mathbf{k})c_{x\sigma}(\mathbf{k}) \right) \qquad (1)$$

where $c_{x\sigma}^{\dagger}(\mathbf{k})$ $(c_{y\sigma}^{\dagger}(\mathbf{k}))$ creates an electron on an 'x' ('y') link of spin σ and momentum $\mathbf{k}$.

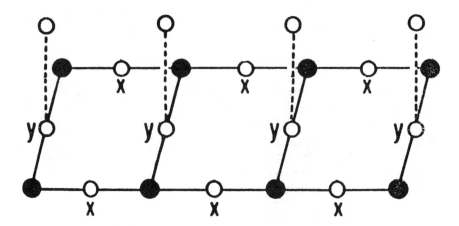

FIGURE 1 The CuO_2 plane. Cu and O are indicated by filled and open circles, respectively. Also pictured are the ordered lines of O in the basal plane of the triplet in high-T_c YBaCuO.

The second key ingredient is the "bond asymmetry term"

$$H_{asym} = -\varepsilon \sum_{\mathbf{k}} \sum_{\sigma} \left(c^{\dagger}_{x\sigma}(\mathbf{k}) c_{x\sigma}(\mathbf{k}) - c^{\dagger}_{y\sigma}(\mathbf{k}) c_{y\sigma}(\mathbf{k}) \right), \tag{2}$$

which favors occupation of 'x' links, and in so doing creates a gap of width 2ε in the oxygen band. Such a term is naturally generated by the orthorhombic distortion present in high-T_c YBaCuO and LaSrCuO, since the various potentials are no longer precisely equal on 'x' and 'y' bonds. An even larger source for a bond asymmetry term in YBaCuO is the presence, in the orthorhombic phase only, of charged ordered lines of oxygen atoms in the middle plane of the triplets, directly above or below just the 'y' links in our picture (Fig. 1). These provide an anisotropic electrostatic contribution of the form (2) (as well as an isotropic piece).

The essence of our results lies in the following simple observation. Consider the familiar BCS formula

$$T_c \sim \omega_c \exp\{-1/\rho_0 g\}, \tag{3}$$

where g is the strength of an attractive 2-body potential, ρ_0 is the average density of states at the Fermi surface μ, and ω_c is some cutoff frequency. Typical low-T_c superconductors have $\rho_0 g \approx \frac{1}{3}$. In the presence of the asymmetry term (2), the electronic energies are $\pm E_{\mathbf{k}}$, with

$$E_{\mathbf{k}} = \left[\left(4t \cos \frac{k_x}{2} \cos \frac{k_y}{2} \right)^2 + \varepsilon^2 \right]^{1/2}. \tag{4}$$

For $|\mu| > \varepsilon$, one calculates

$$\rho_0 \sim \frac{|\mu|}{t\sqrt{\mu^2 - \varepsilon^2}} \, log \left(\frac{t^2}{\mu^2 - \varepsilon^2} \right), \tag{5}$$

and hence

$$T_c(\mu) \sim \omega_c \exp \left\{ -\frac{t\sqrt{\mu^2 - \varepsilon^2}}{g|\mu| \log \left(\frac{t^2}{\mu^2 - \varepsilon^2} \right)} \times f(\mu, \varepsilon) \right\} \tag{6}$$

where the function f depends on details of the potential. Clearly, if $|\mu|$ and ε are tuned experimentally to be close to one another, e.g., by varying chemical composition or the percentage of oxygen defects, T_c can be hugely enhanced.

The logarithmic singularity in Eq. (5) has been noted by many authors;[7] by itself, it proves inadequate in explaining the new generation of superconductors. Note that it survives in the symmetry-restoring limit $\varepsilon \to 0$. The square-root singularity—which does not—is the new feature.

In principle, a square-root singularity in the density of states is not peculiar to two dimensions, but can occur in d dimensions as well, most easily if the Fermi surface at a band edge is a d-dimensional cube. (Eq. (5) is an example of this: the Fermi surface at $\mu = \pm \epsilon$ is the square defined by k_x or $k_y = \pm \pi$.) We now give a physical argument supporting the crucial role of quasi-two-dimensionality. Consider the inevitable perturbations by next-nearest-neighbor hopping terms, which serve to restore some curvature to surfaces of constant energy. Conventional, *round* Fermi surfaces near a band edge at ε are characterized by

$$\rho_0 \sim (|\mu| - \varepsilon)^{\frac{d}{2}-1}, \tag{7}$$

which is constant in 2D but gives a square-root *suppression* at the band edge in 3D. One therefore expects next-nearest-neighbor terms, which interpolate between (5) and (7), to have a *much less deleterious effect* on T_c in 2D than in 3D.**

The bond asymmetry mechanism works in an altogether different way in the 2D tight-binding model that Mattheiss[1] posits as an excellent approximation to his fully three-dimensional band-structure calculations for LaSrCuO and YBaCuO. In this model, the hopping is strictly from copper to nearest-neighbor oxygens in the plane and *vice versa*:

$$H_{hop}^{CuO} = -2t \sum_{k\sigma} \sum_{\alpha=x,y} \left(\cos \frac{k_\alpha}{2} c_{\alpha\sigma}^\dagger(\mathbf{k}) d_\sigma(\mathbf{k}) + \text{h.c.} \right) + \sum_{k\sigma} E_d d_\sigma^\dagger(\mathbf{k}) d_\sigma(\mathbf{k}), \tag{8}$$

where d destroys an electron on the Cu d-shell, and E_d is the energy of the copper level measured from that of the oxygen. One finds two bands with energies

$$\frac{1}{2}E_d \pm \frac{1}{2}\left(E_d^2 + 4t\cos^2\frac{k_x}{2} + 4t\cos^2\frac{k_y}{2} \right)^{1/2}, \tag{9}$$

plus an infinitely narrow band at zero energy corresponding to localized states on the oxygen atoms. To leading order in ε/t, inclusion of the asymmetry term (2) gives the latter band a width 2ε, and a dispersion

$$E'_\mathbf{k} = \varepsilon \frac{\cos^2\frac{k_x}{2} - \cos^2\frac{k_y}{2}}{\cos^2\frac{k_x}{2} + \cos^2\frac{k_y}{2}}. \tag{10}$$

This corresponds to a density of states

$$\rho_0 \sim \frac{\varepsilon}{\varepsilon^2 - \mu^2} \log \frac{1}{|\mu|}\left(\varepsilon + \sqrt{\varepsilon^2 - \mu^2} \right), \tag{11}$$

which has a square-root divergence as before when $\mu \to \pm\varepsilon$.

There are two interesting regimes in which the BCS weak-coupling formula (3), with ρ_0 square-root-enhanced as in (5) or (11), breaks down. For μ close to ε,

---

**This might account for the comparatively low T_c's for the oxide superconductors Ba(Pb,Bi)O, which have a 3D structure. It may also explain the recently reported $T_c \gtrsim 260\,\text{K}$ for a phase of YBaCuO which has a longer 'c' axis, hence is more quasi-two-dimensional.[12]

the nature of the asymptotic expansion of the gap equation for large βt changes completely; instead, one finds power-law behavior, with limiting cases

$$T_c \sim \begin{cases} (g/t)\sqrt{\varepsilon\omega_c} : & \omega_c/2T_c \ll 1 \\ \dfrac{\varepsilon(g/t)^2}{\left(\text{const.}+(g/t)\sqrt{\varepsilon/\omega_c}\right)^2} : & \omega_c/2T_c \gg 1. \end{cases} \tag{12}$$

Numerically, as we shall see in Fig. 2, the changeover from the exponential regime, Eq. (6), to the power-law regime (12) actually cuts off the steep rise in T_c as $|\mu| \to \varepsilon$, preventing it from reaching its "natural" scale of ω_c. Another important consequence of (12) is a *much smaller isotope effect* in this regime; for example, in the model worked out in Sec. II below, we shall find that T_c^{max} effectively scales like $(\omega_c)^{.26}$. Experimentalists should take note that, for fixed g increasing ε increases T_c, so long as $|\mu|$ tracks ε (see Fig. 2).

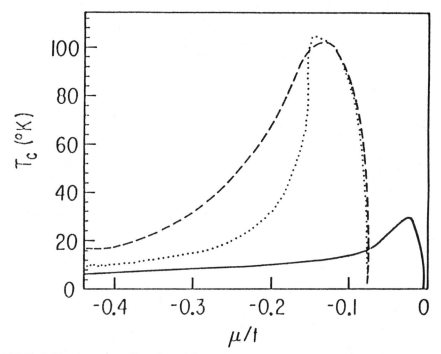

FIGURE 2 T_c *versus* μ. The dotted line corresponds to the phonon exchange model (Sec. II), with $\varepsilon = .1t, \omega_c = .05t, V_0 = 3.6t$. The dashed and solid lines give the analogous result for the static model (Sec. III), with $U_0 = 8t, U_2 = 2t$, and, respectively, $\varepsilon = .1t$ or $\varepsilon = .01t$. We take $t = 5000$ K throughout, corresponding to a band width $8t = 40000$ K. In both models, the coupling constants are sufficiently weak that the corresponding curves for $\varepsilon = 0$ would be hardly visible on this scale.

The second regime occurs when μ passes through ε into the gap. Here one finds that the BCS gap equation can no longer be satisfied, and T_c falls abruptly to zero. As μ enters this regime, our model superconductor makes a sudden transition to a semiconductor—a surprising feature heretofore observed only in the laboratory.

Many experimentalists have remarked on the inhomogeneous nature of high-T_c superconductivity.[8] At the present time, it is not certain whether this is an intrinsic property, or the result of sample inhomogeneity. From the theoretical perspective developed here, it is conceivable that in thermodynamic equilibrium, inhomogenous phases (domains) develop so as to accommodate a given supply of carriers.[9] Thus a fraction p of the material may have a distortion energy $\varepsilon_1 \lesssim |\mu|$ for optimal superconductivity, while the remaining $1 - p$ acquires a value $\varepsilon_2 > |\mu|$, hence has semiconducting properties , in such a way that μ is constant throughout the sample. In contrast to the usual granular superconductors,[10] the value of p, the geometry and size of individual domains, and the values of ε_1 and ε_2 are not given a priori, but are to be adjusted for optimal free energy.

Sections II and III realize the bond asymmetry mechanism for T_c enhancement described above in the context of two specific models, both based (for simplicity) on the hopping term H_{hop}^{xy} given in Eq. (1), but characterized by quite different potential: the first a variant of the usual phonon-mediated weak-coupling BCS, the second relying on attractive static forces of arbitrary origin. We find surprisingly similar, but not identical, results in both models.

II. PHONON EXCHANGE MODEL

We consider the Hamiltonian

$$H = H_{phonon} + H_{hop}^{xy} + H_{asym} - \mu\hat{N},\qquad(13)$$

where $\hat{N}$ counts the total number of electrons. As the charged carriers in this idealized model are assumed to live exclusively on the oxygens, their hopping should be primarily correlated with the vibrations of the nearby Cu atoms, just as copper-centered carriers would have their electron-phonon interaction principally with oxygen vibrational modes.[16] Accordingly, we take[12]

$$H_{phonon} = \sum_{n,m}\left(\frac{p_{n,m}^2}{2M_{Cu}} + \frac{K}{2}\left[(x_{n,m+1} - x_{nm})^2\right.\right.$$

$$+ \left(x_{n+1,m} - x_{nm} + d\sum_\sigma c_{x\sigma}^\dagger\left(n + \frac{1}{2}, m\right)c_{x\sigma}\left(n + \frac{1}{2}, m\right)\right)^2 + (y_{n+1,m} - y_{nm})^2$$

$$+ \left.\left.\left(y_{n,m+1} - y_{nm} + d\sum_\sigma c_{y\sigma}^\dagger\left(n, m + \frac{1}{2}\right)c_{y\sigma}\left(n, m + \frac{1}{2}\right)\right)^2\right]\right).\qquad(14)$$

Eq. (14) is an harmonic-oscillator description of the vibrations of the Cu sublattice; the explicit introduction of an electron-phonon interaction is designed to reduce the equilibrium distance of the bond between adjacent Cu's by an amount d in the presence of an electron on the intervening O atom.

It is convenient to rotate to a frame in which the last three terms in Eq. (13) are diagonal. The appropriate operators are $c_{\pm,\sigma}(\mathbf{k}) = \pm\rho_{\pm}c_{x,\sigma}(\mathbf{k}) + \rho_{\mp}c_{y,\sigma}(\mathbf{k})$, with $\rho_{\pm}(\mathbf{k}) = [(E_{\mathbf{k}} \pm \varepsilon)/2E_{\mathbf{k}}]^{1/2}$ and $E_{\mathbf{k}}$ given in (4). In this basis

$$H_{\text{hop}}^{xy} + H_{\text{asym}} - \mu\hat{N}$$
$$= \sum_{\mathbf{k}\sigma}\left[(E_{\mathbf{k}} - \mu)c_{-,\sigma}^{\dagger}(\mathbf{k})c_{-,\sigma}(\mathbf{k}) - (E_{\mathbf{k}} + \mu)c_{+,\sigma}^{\dagger}(\mathbf{k})c_{+,\sigma}(\mathbf{k})\right]. \quad (15)$$

For $\mu < 0$, it is legitimate to ignore the '$-$' states entirely for temperatures $T \ll 2\varepsilon$, and we shall do so from now on.

In second-order in d, the cross-terms in Eq. (14) allow an electron from the '$+$' band, of energy $-E_{\mathbf{k}}$ slightly below μ, to emit and absorb a phonon and scatter to an energy $-E_{\mathbf{k}'} > \mu$. This results in an energy denominator of the form $E_{\mathbf{k}} - E_{\mathbf{k}'} - \omega_{\mathbf{k}-\mathbf{k}'}$, which can give a large, negative enhancement to the matrix element so long as $|E_{\mathbf{k}} - E_{\mathbf{k}'}|$ is less than some cutoff frequency ω_c. Following BCS, we model the effect of this denominator by restricting $E_{\mathbf{k}}$ and $E_{\mathbf{k}'}$ individually to be within ω_c away from $|\mu|$; in this region the denominator is simply replaced by a negative constant. The net result of these manipulations is an effective interaction term

$$\frac{1}{N}\sum_{\mathbf{k}}\sum_{\mathbf{k}'}V_{\mathbf{k}\mathbf{k}'}c_{+\uparrow}^{\dagger}(\mathbf{k}')c_{+\downarrow}^{\dagger}(-\mathbf{k}')c_{+\downarrow}(-\mathbf{k})c_{+\uparrow}(\mathbf{k}), \quad (16)$$

where $V_{\mathbf{k}\mathbf{k}'} \equiv 0$ except when both $|E_{\mathbf{k}} - |\mu||$ and $|E_{\mathbf{k}'} - |\mu|| < \omega_c$, in which case

$$V_{\mathbf{k}\mathbf{k}'} = -V_0\left[\left(\sin\frac{1}{2}(k_x - k_x')\rho_+(\mathbf{k})\rho_+(\mathbf{k}')\right)^2\right.$$
$$\left. + \left(\sin\frac{1}{2}(k_y - k_y')\rho_-(\mathbf{k})\rho_-(\mathbf{k}')\right)^2\right]. \quad (17)$$

In (16)–(17) we have discarded all terms that vanish in the BCS ground state, and have grouped together all the constants into the positive quantity V_0. The factors of $\rho_{\pm}$ come from substituting $c_{\pm}$ into (14), while the sines result from the fact that the carriers in our model are displaced by half a lattice distance from the Cu sites.

The BCS gap equation follows in the usual way:

$$\Delta(\mathbf{k}) = -\frac{1}{2N}\sum_{\mathbf{k}'}V_{\mathbf{k}\mathbf{k}'}\frac{\Delta(\mathbf{k}')}{\sqrt{(E_{\mathbf{k}'} + \mu)^2 + |\Delta(\mathbf{k}')|^2}}$$
$$\times \tanh\left(\frac{1}{2}\beta\sqrt{(E_{\mathbf{k}'} + \mu)^2 + |\Delta(\mathbf{k}')|^2}\right). \quad (18)$$

$V_{\mathbf{kk'}}$ is a sum of separable potentials, since $\sin^2 \frac{1}{2}(k_x - k'_x) = \frac{1}{2}(1 - \cos k_x \cos k'_x - \sin k_x \sin k'_x)$. Every solution of (18) must therefore be of the form

$$\Delta(\mathbf{k}) = \rho_+(\mathbf{k})^2(\delta_1 + \delta_2 \cos k_x + \delta_3 \sin k_x) + \rho_-(\mathbf{k})^2(\delta_4 + \delta_5 \cos k_y + \delta_6 \sin k_y), \quad (19)$$

where the δ_i's depend only on temperature. Thus the gap varies insignificantly with $\mathbf{k}$, and in general, contains a mixture of partial waves.

The gap equation simplifies at T_c, where the even and odd pieces of $\Delta(\mathbf{k})$ decouple. In fact, since $V_0 > 0$, the odd sector has no nontrivial solution: $\delta_3(T) = \delta_6(T) \equiv 0$. The remaining four δ's satisfy four linear homogenous equations, and T_c can be extracted numerically by requiring the determinant to vanish.

In Fig. 2 we have plotted T_c as a function of μ for parameters chosen to give a maximum T_c of $\gtrsim 100\,\mathrm{K}$. The various regimes discussed in the Introduction are apparent: As μ increases, the rise in T_c is of the general shape predicted by Eq. (6), until $\mu \approx -1.4\varepsilon$, at which point power-law behavior sets in and the curve levels off. There follows a rapid drop into the semiconductor phase at $\mu \approx -.75\varepsilon$, where the gap equation no longer has a (nontrivial) solution. Note that, at the point of maximum T_c, μ is positioned sufficiently far away from the band edge at $-\varepsilon$ to allow for a substantial number of carriers, hence a potentially high supercurrent density. In the high T_c regime, $\delta_2/\delta_1 \approx .4$, while δ_4/δ_1 and $\delta_5/\delta_1 < .1$. Thus the Cooper pairs are formed primarily as s-waves with a significant d-wave admixture; the fact that $\delta_5 < \delta_2$ implies anisotropy within the basal plane.

In our model, Eq. (14), ω_c explicitly depends only on copper mass, not on oxygen mass. For the parameters chosen in Fig. 2, in the region of maximum T_c, we have calculated $T_c \propto (\omega_c)^{.26}$, hence a very small Cu isotope effect, as compared with $T_c \propto \omega_c$ for μ far from ε. Note, though, that the value of the exponent is highly parameter-dependent (see Eq. (12)).

III. STATIC MODEL

We now turn to a general class of models characterized by a static (*i.e.*, non-velocity-dependent) potential, which, for convenience, shall be assumed to connect only 'x' carriers to 'x' carriers, and 'y' carriers to 'y' carriers:

$$\mathsf{H} = \mathsf{H}_{\mathrm{static}} + \mathsf{H}_{\mathrm{hop}}^{\mathrm{xy}} + \mathsf{H}_{\mathrm{asym}} - \mu \hat{N}, \quad (20)$$

where

$$\mathsf{H}_{\mathrm{static}} = \frac{1}{n} \sum_{\mathbf{k},\mathbf{k'}} \sum_{\alpha=x,y} V_{\mathbf{kk'}}^{\alpha} c_{\alpha\uparrow}^{\dagger}(\mathbf{k'}) c_{\alpha\downarrow}^{\dagger}(-\mathbf{k'}) c_{\alpha\downarrow}(-\mathbf{k}) c_{\alpha\uparrow}(\mathbf{k}). \quad (21)$$

Making the usual BCS quadratic approximation to this potential, which is quartic in field operators, we can rewrite H as follows:

$$\mathsf{H} = \sum_{\mathbf{k}} \left(\sum_{\alpha=x,y} \Delta_\alpha(\mathbf{k}) < c^\dagger_{\alpha\uparrow}(\mathbf{k}) c^\dagger_{\alpha\downarrow}(-\mathbf{k}) > -2\mu \right)$$

$$+ \sum_{\mathbf{k}} \Psi \begin{pmatrix} -\mu - \varepsilon & 0 & -t_{\mathbf{k}} & -\Delta_x(\mathbf{k}) \\ 0 & \mu - \varepsilon & -\Delta_y^*(\mathbf{k}) & t_{\mathbf{k}} \\ -t_{\mathbf{k}} & -\Delta_y(\mathbf{k}) & -\mu + \varepsilon & 0 \\ -\Delta_x^*(\mathbf{k}) & t_{\mathbf{k}} & 0 & \mu + \varepsilon \end{pmatrix} \Psi^\dagger, \qquad (22)$$

where Δ_x and Δ_y are the gap functions, $t_{\mathbf{k}} = 4t \cos\frac{k_x}{2} \cos\frac{k_y}{2}$, and

$$\Psi = \left(c^\dagger_{x\uparrow}(\mathbf{k}), c_{y\downarrow}(-\mathbf{k}), c^\dagger_{y\uparrow}(\mathbf{k}), c_{x\downarrow}(-\mathbf{k}) \right).$$

It is straightforward to find the Bogoliubov transformation that diagonalizes the matrix, but for our present purposes the eigenvalues $\{\mathsf{E}_{\mathbf{k}}^\pm, -\mathsf{E}_{\mathbf{k}}^\pm\}$ suffice. They are given by

$$(\mathsf{E}_{\mathbf{k}}^\pm)^2 = t_{\mathbf{k}}^2 + \varepsilon^2 + \mu^2 + \frac{1}{2}(|\Delta_x|^2 + |\Delta_y|^2)$$

$$\pm \frac{1}{2}\sqrt{16\mu^2 t_{\mathbf{k}}^2 + 4t_{\mathbf{k}}^2 |\Delta_x - \Delta_y|^2 + (4\mu\varepsilon + |\Delta_x|^2 - |\Delta_y|^2)^2}. \qquad (23)$$

Using Feynman's theorem, we can express the two coupled gap equations for $\Delta_\alpha(\mathbf{k})$ in an especially compact form:

$$\Delta_\alpha(\mathbf{k}) = -\frac{1}{N} \sum_{\mathbf{k}'} \sum_{\tau=\pm 1} V^\alpha_{\mathbf{k},\mathbf{k}'} \frac{\partial \mathsf{E}_{\mathbf{k}'}^\tau}{\partial \Delta_\alpha^*} \tanh \frac{1}{2}\beta \mathsf{E}_{\mathbf{k}'}^\tau, \qquad \alpha = x, y. \qquad (24)$$

The recent literature abounds with suitable candidates for an attractive potential. As a concrete example, we have examined

$$V^x_{\mathbf{k}\mathbf{k}'} = U_0 - U_2 \cos(k_x - k_x'), \qquad V^y_{\mathbf{k}\mathbf{k}'} = U_0 - U_2 \cos(k_y - k_y'). \qquad (25)$$

Here U_0 and U_2 are positive constants with U_0 representing Coulomb repulsion. The anisotropy built into the cosine terms in V^x and V^y is consistent with the topology of the lattice: two 'x' carriers displaced vertically or horizontally from one another are separated by vacuum or by a Cu atom, respectively.

The solutions for the potential (25) are all of the form

$$\Delta_x(\mathbf{k}) = \delta_1 + \delta_2 \cos k_x + \delta_3 \sin k_x, \qquad \Delta_y(\mathbf{k}) = \delta_4 + \delta_5 \cos k_y + \delta_6 \sin k_y. \qquad (26)$$

In this model, there *is* an odd solution, thanks to the sign of U_2. However, for all ranges of parameters that we have considered, the even solution has the higher T_c, and hence lower free energy, at least for temperatures near T_c. We therefore set $\delta_3(T) = \delta_6(T) \equiv 0$, although we should mention the intriguing possibility of a new phase transition at a lower temperature in which these components, which are associated with p-wave spin-1 Cooper pairs, "turn on," perhaps discontinuously.

Fig. 2 also depicts T_c *versus* μ as we have computed it for the even solution to (24), again with a choice of parameters that gives a maximum $T_c \gtrsim 100\,\mathrm{K}$. (Also shown is the result of reducing ε by a factor of 10.) The most surprising feature of the curve is that, despite the radically different potential, it is quite similar to its counterpart in the phonon-exchange model. In the present case, however, the Cooper pairs are primarily d-wave rather than s-wave: $\delta_1/\delta_2 \approx .5$, $\delta_4/\delta_2 \approx -.15$, $\delta_5/\delta_2 \approx -.1$ in the vicinity of maximum T_c.

Somewhat more details are now available.[13] The extension of our theory to the Eliashberg formalism is presently underway.[14]

ACKNOWLEDGEMENTS

The work was supported in part by ONR contract N00014–86–K–0710, DOE contract DE–AC02–82ER–40073, and by an Enrico Fermi Fellowship. We are indebted to T. Banks, J. Crow, A. Kapitulnik, M. Karliner, K. Levin, O. Symko, A. Virkar, M. Weinstein and especially M. Peskin and A. Szpilka for valuable information and insightful conversations. M.M. gratefully acknowledges the hospitality of the SLAC theory group where much of this work was completed.

REFERENCES

1. L. F. Mattheiss, *Phys. Rev. Lett.*, **58**, 1028 (1987); L. F. Mattheiss and D. R. Hamann, *Sol. St. Comm.*, **63**, 395 (1987).
2. J. Yu, A. Freeman, J.-H. Xu, *Phys. Rev. Lett.*, **58**, 1035 (1987); S. Massidda, J. Yu, A. Freeman, D. Koelling, *Phys. Lett. A*, **122**, 198 (1987), and this volume.
3. See especially P. P. Freitas and T. S. Plaskett, "High temperature order-disorder phase transition in the $Y_1Ba_2Cu_3O_{6+\delta}$ superconductor observed by electrical resistivity measurements," *Phys. Rev. B*, **36**, 5723 (1987); R. M. Fleming, *et al.*, *Phys. Rev. B*, **35**, 7191 (1987); and I. Schuller, *et al.*, *Sol. St. Comm.*, **63**, 385 (1987).
4. B. Batlogg, *et al.*, *Phys. Rev. Lett.*, **58**, 2333 (1987), and *ibid*, **59**, 912 (1987); L. C. Bourne, *et al.*, *ibid.*, **58**, 2337 (1987); T. Faltens, *et al.*, *ibid.* **59**, 915 (1987).
5. J. Bardeen, L. Cooper, and R. Schrieffer, *Phys. Rev.*, **108**, 1175 (1957).
6. N. P. Ong *et al.*, *Phys. Rev. B*, **35**, 8807 (1987).

7. See, *e.g.*, Ref. 1 and 2, and also J. D. Jorgensen, *et al.*, *Phys. Rev. Lett.*, **58**, 1024 (1987); J. E. Hirsch and D. J. Scalapino, *Phys. Rev. Lett.*, **56**, 2732 (1986); P. A. Lee and N. Read, *Phys. Rev. Lett.*, **58**, 2691 (1987)

8. X. Cai, R. Joint, D. Larbalestier, *Phys. Rev. Lett.*, **58**, 2798 (1987).

9. See also M. Weinstein, (1987) SLAC-PUB 4272.

10. See, *e.g.*, C. Ebner and D. Stroud, *Phys. Rev. B*, **31**, 165 (1985).

11. W. Weber, *Phys. Rev. Lett.*, **58**, 1371 (1987).

12. D. C. Mattis, *Phys. Rev. B*, **36**, 3933 (1987).

13. D. C. Mattis and M. P. Mattis, *Phys. Rev. Lett.*, **59**, 2780 (1987).

14. M. P. Mattis, P. Arnold, and D. C. Mattis, to be submitted.

J. Ruvalds
Physics Department
University of Virginia
Charlottesville, Virginia 22901

Plasmons and High-Temperature Superconductivity in Alloys of Copper Oxides[†]

The two-dimensional character of the electronic structure of M-N-Cu-O alloys, with M =La, Y and N =Ba, Sm, ..., is shown to favor the formation of "acoustic" plasmons at energies above the acoustic phonons. Providing that the Fermi energy intersects a small pocket of electrons (or holes) in addition to the expected occupation of a primary electron (or hole) band, the plasma oscillations of the secondary charge carriers may provide a mechanism for room-temperature superconductivity.

The recent discovery of alloys which remain superconducting at high temperatures has spurred a remarkable surge of interest and research activity. The initial low-key report[1] of a superconducting transition near 30 K in La-Ba-Cu-O compounds set the stage for a rapidly escalating series of measurements and reports of new materials such as Y-Ba-Cu-O with transition temperatures of $T_c \simeq 90\,\mathrm{K}$.[2] Superconducting features at even higher temperatures have been reported in *The New York Times*, and the quest for a room-temperature superconductor has been rejuvenated with intense and spirited rivalry.

[†]Reprinted from *Phys. Rev. B*, **35**, 8869 (1987).

Prior to these developments, the superconducting properties of most metals were explained by the BCS theory,[3] which invokes a phonon-mediated interaction between electrons to form superconducting pairs under certain circumstances. A slightly modified form of the weak coupling result for the transition temperature can be written as

$$T_c \simeq 0.7\theta_D \exp\left(-\frac{1+\lambda}{\lambda-\mu^*}\right), \tag{1}$$

where θ_D is a Debye frequency representing the lattice vibrations, λ is the electron-phonon coupling parameter, and the modified Coulomb repulsion μ^* is defined by

$$\mu^* = \frac{\mu}{1+\mu\ln(E_F/\theta_D)}. \tag{2}$$

In typical metals the phonon frequencies are much smaller than the Fermi energy E_F, and thus the effect of the average Coulomb repulsion μ between electrons is reduced to a small correction of order $\mu^* \simeq 0.1$. Hence if the electron-phonon coupling λ can be of order unity, and if $\theta_D \simeq 300$ K, it is reasonable to expect that $T_c \simeq 20$ K. However, considerably higher transition temperatures are difficult to envision for the phonon spectra of ordinary metals.

An exceptional case which has long been regarded as a candidate for a room-temperature superconductor is metallic hydrogen. By virtue of its light ion mass the hydrogen phonon vibrations may extend to very high values, say $\theta_D \simeq 3000$ K, and thereby enhance the T_c prospects as seen in Eq. 1. Calculations[4] of the corresponding λ and μ^* parameters for metallic hydrogen support the hope of $T_c \simeq 200$ K, and thus lend impetus to the challenge of creating a metallic hydrogen state in the laboratory.

Other excitations with much higher frequencies, such as excitons and plasmons have also been examined in the past, but they suffer a dual handicap. On the one hand, if the equivalent "Debye" energy θ is too large, say $\theta \simeq E_F$, then the Coulomb repulsion between electrons is not reduced as much as in the phonon case: a glance at Eq. (2) illustrates the most direct consequence of the high-energy modes. Furthermore, the Migdal theorem becomes doubtful and a much more detailed analysis of the dielectric screening is required. Yet another difficulty arises for very high-frequency excitations from the solutions of the Eliashberg equations for superconducting pairing. These yield an expression for the boson-mediated electron pairing parameter.

$$\lambda = 2\int \frac{\alpha^2 F(\omega)}{\omega} d\omega, \tag{3}$$

where $F(\omega)$ is the boson density of states and α^2 represents the coupling strength between electrons and bosons. From this relation, it appears that λ is reduced in the case of commonplace high-frequency bosons. Hence, elevating the boson energy too far has the combined destructive influence in reducing the electron-electron attraction λ, increasing their repulsion μ^*, and thereby dramatically lowering the exponential term in Eq. (1); the result is a negation of the positive influence on T_c achieved by a larger prefactor θ.

Incidentally, the above considerations provide part of the motivation for the commonly used soft phonon mechanism for enhancing T_c, in the sense that very low-energy phonons may enhance the λ parameter sufficiently to overcome a smaller prefactor θ in that case.

An alternate mechanism for electron pairing mediated by intermediate energy acoustic plasmons in transition metals was originally proposed by Fröhlich.[5] A review of research efforts to establish the possible existence of these modes in metals is presented in Ref. 6.

The recently discovered high-T_c superconductors are notable for an extreme sensitivity to alloy composition. For example, La_2CuO_4 is not superconducting, whereas $La_{2-x}\,Ba_xCu_4$ has $T_c \simeq 40\,K$ under pressure with $x \simeq 0.2$, and $(Y_{1-x}Ba_x)_2CuO_4$ has $T_c \simeq 90\,K$ providing $x \simeq 0.4$.[2] This unusual behavior provides the principal clue to our investigation, especially since bulk properties such as the averaged phonon spectrum or the ordinary plasmon response are not expected to vary dramatically over such a small impurity concentration range.

We propose that the superconductivity of the La-M-Cu-O alloys may be attributed to low-energy plasmon excitations extending to corresponding values of $\theta_{pl} \simeq 3000\,K$. The prospects for creating such modes are greatly enhanced by the two-dimensional nature of the electronic energy spectrum which gives rise to an "acoustic plasmon branch", in the sense that the plasmon energy behaves as

$$\omega_q \simeq (aq + bq^2)^{1/2}, \tag{4}$$

where the coefficients a and b are determined by the dielectric screening and the branch remains well defined outside the electron-hole continuum up to a cutoff momentum q_c.[7,8] However, we emphasize that a single branch, appropriate to La_2CuO_4 for example, is not a likely candidate for inducing superconducting electron pairing because of its very high frequency ($\theta \sim 2\,eV$) and the arguments cited above. Hence, we are led to examine a secondary, lower energy, plasmon branch which may be created by proper alloying.

A likely situation for the formation of the low-energy plasmon models is illustrated in Fig. 1 where the representation of the electronic structure near the Fermi energy has been patterned after the current band-structure results[9,10] for La_2CuO_4. The key feature of these electron bands is the two-dimensional nature of the electron dynamics. Thus, within the plane of motion, we may approximate the energies as $E_h = (k_x^2 + k_y^2)/2m_h^*$ and $E_\ell = (k_x^2 + k_y^2)/2m_\ell^*$, where the effective masses m_ℓ^* and m_h^* refer to the curvature of the ℓ and h bands, respectively. The two-dimensional character of the bands greatly favors the criterion for creating undamped plasmon modes at intermediate energies with a dispersion relation of Eq. (4), which is reminiscent of acoustic modes with their $\omega \simeq cq$. If the Fermi energy intersects only the E_ℓ band, as shown in Fig. 1 for the La_2CuO_4 compound, then a single plasmon branch will appear. However, the interesting possibility regarding high-T_c superconductivity is the situation where the Fermi energy intersects both bands, thus

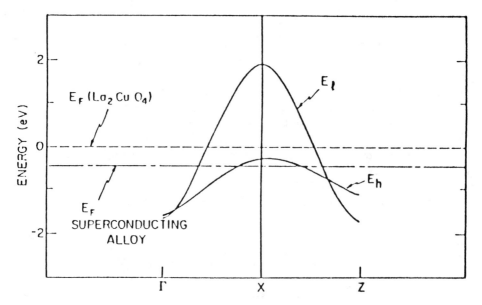

FIGURE 1 Representation of the electron energy as a function of momentum. The shape of E_ℓ and E_h, as well as the Fermi energy E_F, for La_2CuO_4 is patterned after the band-structure calculations of Refs. 9 and 10. The influence of alloying may lower the Fermi energy as shown by the dot-dash line intersecting both bands.

creating a small pocket of h charge carriers which can oscillate at intermediate energies and mediate the pairing interaction of the ℓ electron (or hole) states.

The plasmon spectrum for the two-band case follows readily from the dielectric function $\epsilon(q,\omega)$ in two dimensions. A convenient choice of dimensionless variables is to measure wave vectors in units of k_F and frequencies in units of $2E_F/\hbar$. Then, introducing $\alpha \equiv 0.22r_s$, where r_s is the usual average electron spacing, and the variables $v_\pm = \pm\omega/k - k/2$, the dielectric function for a single band can be evaluated to find[7,8]

$$\text{Re}\,\epsilon(q,\omega) =$$
$$1 + \frac{2\pi\alpha}{k^2}\left[k + \text{sgn}(v_+)\theta(v_+^2 - 1)(v_+^2 - 1)^{1/2} + \text{sgn}(v_-)\theta(v_-^2 - 1)(v_-^2 - 1)^{1/2}\right], \quad (5a)$$

and

$$\text{Im}\,\epsilon(q,\omega) = \frac{2\pi\alpha}{k^2}\left[\theta(1 - v_+^2)(1 - v_+^2)^{1/2} - \theta(1 - v_-^2)(1 - v_-^2)^{1/2}\right]. \quad (5b)$$

The total dielectric function is $\epsilon_{\text{total}} = \epsilon_\ell + \epsilon_h$, and the individual ℓ and h contributions naturally include their respective screening parameters r_s, E_F, and k_F. The plasmon dispersion in this two-dimensional case is quite different from the usual three-dimensional "optic" plasmon counterpart. From Eq. (5), even a single electron band will generate an "acoustic" plasmon mode[7,8] with the energy spectrum $\omega_{\text{pl}} = (aq + bq^2)^{1/2}$, where $a \equiv \pi\alpha$ and $b = 0.75$ in these units. The resulting two-band spectrum can be readily calculated from Eq. (5), using the band-structure model appropriate to alloys of La-M-Cu-O. The results are shown in Fig. 2, where the two shaded regions of the ℓ and h electron-hole continuum indicate the values of ω and q where Landau damping of the plasmons would be possible by virtue of the nonvanishing imaginary part $\mathrm{Im}\,\epsilon(q,\omega)$. The small pocket of h charge carriers shifts the ℓ-plasmon branch somewhat, but this higher-frequency mode is expected to persist whether or not the Fermi energy intersects both ℓ and h bands, and therefore may be present in nonsuperconducting alloys in the La-M-Cu-O series. However, the lower-energy h-plasmon branch becomes a candidate for a supercon-ductivity mechanism only if the Fermi energy intersects the E_h band as well. In that case, the screening of the h electrons (or holes) by the primary ℓ carriers tends to shift the "acoustic" h branch to higher frequencies, as illustrated in Fig. 2. Since this h branch of the plasmon spectrum may have a corresponding "Debye" energy of $\theta^h_{\mathrm{pl}} \sim 3000$ K, it has many of the desirable features which are anticipated for a plasmon-mediated mechanism for high temperature superconductivity.

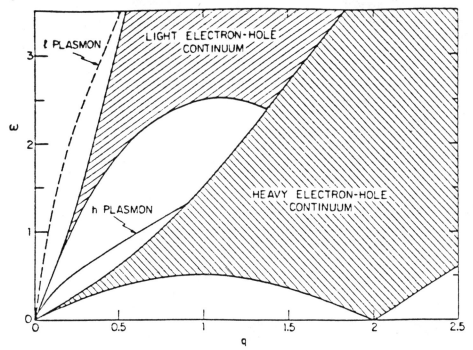

FIGURE 2 Excitation spectrum proposed for certain alloys of the La-M-Cu-O series. The energy ω and the momentum q are scaled to the Frmi energy measured from the top of the E_h secondary band pocket. Thus the lower "acoustic" h plasmon may extend to an energy of order $\omega^h_{\mathrm{pl}} \simeq 0.25$ eV $= 3000$ K. The primary l plasmon is expected to extend to much higher energies, $\omega^l_{\mathrm{pl}} \simeq 2$ eV, and should be present in La$_2$CuO$_4$, while the lower-energy h-plasmon branch is predicted only for the high-T_c superconducting alloys.

Finally, we estimate the corresponding electron pairing parameter λ and μ^* corresponding to the "acoustic" readily from the dielectric function $\epsilon(q,\omega)$ in two dimensions. plasmon exchange process. Our analysis resembles the calculations for metallic hydrogen.[4] For the μ^* defined in Eq. (2) we have the standard definition of the averaged Coulomb repulsion.

$$\mu = N(0) \int_0^{2q_F} \frac{v_q q \, dq}{q_F^2 \epsilon(q,0)} , \qquad (6)$$

where v_q is the Coulomb interaction, $N(0)$ is the density of states at the Fermi energy, and $\epsilon(q,0)$ is the static limit of Eq. (5). For parameters appropriate to the placement of the Fermi energy in Fig. 1 (for the superconducting alloy) we estimate $\mu = 4.6$ and $\mu^* \sim 0.4$. As long as the h-plasmon energy θ_{pl}^h remains below 3000 K, the μ^* parameter is thus reasonably small. Computation of the electron-plasmon interaction follows the Eliashberg equation solution of Eq. (3) with $\alpha^2 = N(0)\langle qv^2(q)\rangle$, and $F(\omega)$ is the density of states for the h-plasmon branch. Here a secondary benefit of the two-dimensional electron dynamics becomes evident because the restricted dimensionality enhances the phonon density of states at lower frequencies and thus enhances λ. The density of states $F(\omega)$ is shown as a function of the dimensionless frequency in Fig. 3 for the h-plasmon dispersion of Eq. (4). For comparison, a truly acoustic-plasmon branch, with $a = 0$, would give $F(\omega) \propto \omega$ in two dimensions and such a pristine mode would increase λ by a factor of 2 over the present case. It is worth mentioning that the three-dimensional analog would be $F_{3D}(\omega) \propto \omega^2$ which yields considerably smaller values. For the case depicted in the figures in this paper we thus estimate $\lambda \simeq 1.4$ and therefore $T_c \simeq 200$ K. The transition temperature is very sensitive to changes in the plasmon frequency θ_{pl}^h. The current model gives the following correlations: for $\theta_{\text{pl}}^h = 1800$ K, $\lambda = 1.5$, $\mu^* = 0.36$, and $T_c = 107$ K; for $\theta_{\text{pl}}^h = 1200$ K, $\lambda = 1.4$, $\mu^* = 0.30$, and $T_c = 60$ K; for $\theta_{\text{pl}}^h = 600$ K, $\lambda = 0.76$, $\mu^* = 0.26$, and $T_c = 32$ K. Hence T_c is strongly related to the position of the Fermi energy and this novel feature suggests interesting experimental tests which may probe the validity of the two-band plasmon mechanism proposed here.

It should be noted that the above estimate of T_c is crude and should be examined independently by including exchange and correlation corrections to the electronic screening as well as more sophisticated analysis of the Eliashberg equation.[11]

Experimental probes of these predicted low-energy plasmon modes in La-M-Cu-O and Y-Ba-Cu-O superconductors may elucidate the optimum conditions for achieving higher transition temperatures. Of course their existence should be established first, and possible measurements include electron tunneling spectroscopy, electron loss, infrared absorption, and high-frequency acoustic attenuation studies. Neutron scattering may indirectly reveal structure in those optical phonons which intersect the plasmon branch.

A limitation on the plasmon mechanism is imposed by the electron scattering rate which may be influenced by impurities, lattice defects, phonon scattering, and possibly by other means. Thus, the intrinsic plasmon damping will be sensitive to

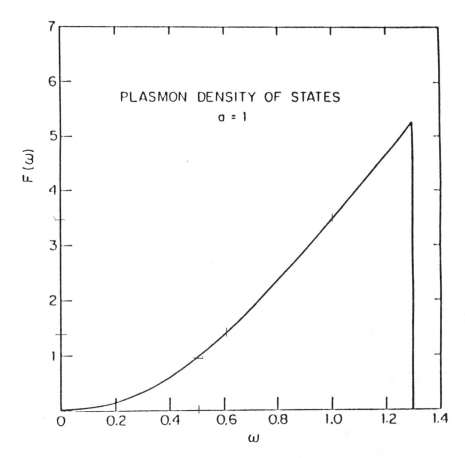

FIGURE 3 The plasmon density of states $F(\omega)$ shown for the two-dimensional dispersion relation using $a = 1$ and $b = 1$. By comparison, two-dimensional acoustic phonons would have $F(\omega) \propto \omega$.

crystal disorder and alloy composition. Also, alloying may shift the Fermi energy or the relative separation of the energy bands in such a way as to allow decay of the lower branch into the electron-hole continuum.

The present results indicate that other materials with two-dimensional electronic structure are likely candidates for high-temperature superconductivity, providing that two appropriate energy bands coexist near the Fermi energy. Naturally, the three-dimensional electron plasma considered originally by Fröhlich[5] and others[6] should also be reexamined in view of the recent discoveries of high-T_c alloys. In this connection, it is worth mentioning that the anomalous concentration dependence of T_c in the oxides referred to as tungsten bronzes, i.e., $M_x WO_3$, has been attributed to an acoustic-plasmon mechanism.[12]

Our conclusions will hopefully serve to stimulate cooperative efforts by experts in band calculations and many-body theorists. The electronic structure provides a valuable guide to the choice of materials which may exhibit acoustic-plasmon oscillations, and their role in superconducting electron pairing merits further study in regard to higher-order screening corrections. Finally, strong coupling solutions of the Eliashberg equations should be examined to ascertain the credibility of the plasmon mechanism as a candidate for room-temperature superconductivity.

After this manuscript was submitted, supporting evidence for the two-band structure envisioned here was obtained for the Y-Ba-Cu-O superconductors from the Mattheiss band-structure calculation[13] and resonant photoemission spectroscopy data.[14]

ACKNOWLEDGEMENTS

It is a pleasure to acknowledge stimulating discussions with H. Gutfreund, A. K. Rajagopal, Z. Tešanović, S. A. Wolf, and several colleagues at Harvard. This research was supported by the U.S. Department of Energy Grant No. DEF605-84-ER45113.

REFERENCES

1. T. G. Bednorz and K. A. Müller, *Z. Phys. B*, **64**, 189 (1986)
2. M. K. Wu, J. R. Ashburn, C. J. Torng, P. H. Hor, R. L. Meng, L. Gao, Z. J. Huang, Y. Q. Wang, and C. W. Chu, *Phys. Rev. Lett.*, **58**, 908 (1987).
3. J. Bardeen, L. N. Cooper, and J. R. Schrieffer, *Phys. Rev.*, **106** 162 (1957).
4. N. Ashcroft, *Phys. Rev. Lett.*, **21**, 1748 (1968); a more recent discussion of superconducting hydrogen in liquid form is in J. E. Jaffe and N. W. Ashcroft, *Phys. Rev. B*, **23**, 6176 (1981).
5. H. Fröhlich, *J. Phys. C*, **1** 544 (1968).
6. J. Ruvalds, *Adv. Phys.*, **30**, 677 (1981).
7. F. Stern, *Phys. Rev. Lett.*, **18**, 546 (1967).
8. A. Czachor, A. Holas, S. R. Sharma, and K. S. Singwi, *Phys. Rev. B*, **25**, 2144 (1982).
9. L. F. Mattheiss, *Phys. Rev. Lett.*, **58**, 1028 (1987).
10. Jaeyun Yu, A. J. Freeman, and J. H. Xu, *Phys. Rev. Lett.*, **58**, 1035 (1987).
11. M. Grabowski and L. J. Sham, *Phys. Rev. B*, **29**, 6132 (1984).
12. L. M. Kahn and J. Ruvalds, *Phys. Rev. B*, **19**, 5652 (1979).
13. L. F. Mattheiss and D. R. Hamann, unpublished.
14. R. L. Kurtz, R. L. Stockbaur, D. Mueller, A. Shih, L. E. Toth, M. Osofsky, and S. A. Wolf, *Phys. Rev. B*, **35**, 8818 (1987).

final.

D. J. Scalapino, R. T. Scalettar
Department of Physics
University of California
Santa Barbara, California 93106

and

N. E. Bickers
Institute for Theoretical Physics
University of California
Santa Barbara, CA 93106

Polaron Effects in High-T_c Oxide Superconductors[†]

We argue that measurements of the specific heat, static susceptibility, dc and ac conductivity, and thermopower are consistent with a polaron model for normal-state conduction in the high-T_c superconductors $La_{2-x}Sr_xCuO_{4-y}$ and $YBa_2Cu_3O_{7-y}$. Strong electron-phonon coupling is also suggested by measurements of elastic constants and thermal Debye-Waller factors. We investigate a model for polaron formation in these systems and a possible mechanism for polaron superconductivity.

The discovery of high-temperature superconductivity[1,2] in the layered perovskite systems $La_{2-x}Sr_xCuO_{4-y}$ and $YBa_2Cu_3O_{7-y}$ has stimulated new interest in unconventional pairing mechanisms. In this paper, we argue that a number of normal-state experiments are consistent with conduction by small polarons. We investigate a polaron pairing mechanism[3] which could lead to superconductivity and argue that the resulting transition temperature may be reasonably high.

[†]Reprinted from *Proceedings of the International Conference on Novel Mechanisms of Superconductivity*, ed. S. E. Wolf and V. Z. Kresin, Plenum, New York, 1987.

Measurements of elastic constants[4,5] and oxygen Debye-Waller factors[6] suggest that the electron-lattice interaction plays an important role in the new systems. In particular, measurements of the sound velocity,[4] ultrasonic attenuation,[4] and Young's modulus[5] indicate that the lattice in $La_{2-x}Sr_xCuO_{4-y}$ is anomalously soft in the temperature range from 200 K down to the superconducting transition. Furthermore, a strong electron-lattice interaction would provide a natural link between the new layered perovskites and the 13 K perovskite superconductor[7] $BaPb_{1-x}Bi_xO_3$: magnetic interactions probably cannot provide such a link, since $BaPb_{1-x}Bi_xO_3$ contains no transition elements.

A number of experiments seem consistent with conduction by small polarons, rather than by electrons weakly coupled to phonons. First of all, measurements of the normal-state specific heat coefficient and magnetic susceptibility[8-11] show that the high-T_c superconductors are strongly-coupled systems. While the Sommerfeld χ/γ ratio appears to be of order unity,[9] the measured thermodynamic density of states is enhanced by an order of magnitude over values from band structure.[12,13] So large a discrepancy probably cannot be explained in terms of a conventional $(1+\lambda)$-enhancement from weak-coupling phonons. (On the other hand, the approximate equality of χ and γ argues against some magnetic mechanisms, which predict a Stoner-like enhancement of the susceptibility.) One explanation for the uniform enhancement of χ and γ is band narrowing by small-polaron formation.

A second argument against the conventional electron-phonon mechanism is the absence of an oxygen isotope effect[14,15] in $YBa_2Cu_3O_{7-y}$. While it is possible to envisage phonon modes which lead to superconductivity yet do not involve large-scale oxygen motion, the coupling to such modes is believed to be weak.[16] The polaron-pairing mechanism investigated below exhibits an anomalous isotope effect, whose details remain to be investigated.

Additional evidence for polaron formation in the normal state is provided by recent measurements of the frequency-dependent electrical conductivity in single crystals of $La_{2-x}Sr_xCuO_{4-y}$.[17] The conductivity $\mathrm{Re}\,\sigma(\omega)$ within the CuO_2 planes exhibits the following qualitative features: (a) a prominent peak near 2 eV, (b) a peak at approximately 0.5 eV, which grows in intensity with increased doping, and (c) a narrow Drude-like feature at zero energy. The integrated weight of the Drude peak is estimated at only 5% of the total oscillator strength. Experiments on polycrystalline samples of $YBa_2Cu_3O_{7-y}$ also suggest the presence of a high-energy peak in $\mathrm{Re}\,\sigma(\omega)$ and drastically reduced weight in the Drude peak. These measurements are consistent with conduction by small polarons:[18] when a polaron hops from site i to $i+1$ (see Fig. 1), it alters the potential energy surface for the local ionic coordinates. A ground-state to ground-state transition is exponentially suppressed by the Franck-Condon factor $\exp(-E_p/\omega_0)$, where E_p is the polaron binding energy and ω_0 is the characteristic phonon frequency. These "zero-phonon" hopping processes lead to formation of a polaron band, whose width is narrowed from the original electronic bandwidth by the factor $\exp(-E_p/\omega_0)$. Scattering processes within this narrow band can account for the Drude peak in the optical conductivity: since the

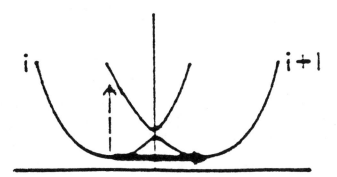

FIGURE 1 Schematic local ion potential for a polaron at sites i and $i+1$. When the polaron hops, the background ions may undergo a "zero-phonon" transition (solid arrow) or, more likely, be left in an excited state (dashed arrow). Only the first process contributes to the formation of a coherent polaron band.

probability for hopping without phonon emission is small, the weight in this feature is expected to be only a small fraction of the total oscillator strength.

On the other hand, the most favorable hopping process is one in which the local ionic coordinates do not rearrange (see Fig. 1). In the simplest approximation, such a process requires an excitation energy of order $2Ep$.[18] A peak is expected in the optical conductivity at this energy. Since the hopping process requires that an unoccupied site be adjacent to an occupied site, the peak should scale with $x(1-x)$, where x is the polaron concentration. In the scenario discussed below, x is just the divalent dopant concentration in $La_{2-x}M_xCuO_{4-y}$. Experimental results are consistent with this dependence in the low doping limit.[17] The value of the dc conductivity is limited by impurity scattering and need not correlate with the height of the peak at energy $2Ep$.[19]

Note that the polaron bandwidth D_P is drastically reduced from the rigid-lattice value. For a two-dimensional tight-binding model with hopping matrix element $\tilde{t}$, the rigid-lattice bandwidth is just $8\tilde{t}$. Assuming $E_P = 0.25$ eV (the rough value indicated by optical measurements in $La_{2-x}Sr_xCuO_{4-y}$), $\omega_0 = 0.1$ eV, and $\tilde{t} = 0.1$ eV gives

$$m^*/m = e^{E_P/\omega_0} = 12$$
$$D_P = 8\tilde{t}(m/m^*) = 800\,\text{K}. \tag{1a}$$

For low dopant concentration x, the polaron Fermi energy is roughly

$$\epsilon_F = \frac{1}{2}xD_P = x(400\,\text{K}). \tag{1b}$$

For such small band energies, unusual thermal effects are possible. In particular, at temperatures of order of the bandwidth, the electrical resistivity due to impurity scattering becomes linear in T, rather than constant. Approximate linearity may persist down to temperatures of order one-tenth the bandwidth.[20] A nearly linear resistivity above T_c has been observed in both $La_{2-x}Sr_xCuO_{4-y}$ and $YBa_2Cu_3O_{7-y}$.[21]

The measured thermopower $S(T)$ of $La_{2-x}Sr_xCuO_{4-y}$ also seems inconsistent with conduction by nearly free carriers, but consistent with polaron conduction.[22] As shown by Cooper, et al.,[22] the thermopower in $La_{2-x}Sr_xCuO_{4-y}$ and $La_{2-x}Ba_xCuO_{4-y}$ is large (of order 100–300 $\mu V/K$ for low doping levels) and nearly constant with temperature in the range 100−−300 K. Further, the high-temperature thermopower is well represented by the formula[18,23]

$$S = -\frac{k_B}{|e|}\ln 2 - \frac{k_B}{|e|}\ln\frac{x}{1-x} \qquad (2)$$

for x between 0 and 0.15. This is the appropriate expression for conduction by spin$-\frac{1}{2}$ polarons with no double site occupancy (the ln 2 term represents the transport of spin entropy).

Transport and thermal properties may be modeled qualitatively by assuming that conduction occurs in an *uncorrelated* narrow band. The presence of correlations to inhibit double site occupancy is quantitatively important: for example, in the absence of correlations, the high-temperature thermopower becomes

$$S = \frac{k_B}{|e|}\ln\frac{x}{2-x}. \qquad (3)$$

Nevertheless, the present calculation for an uncorrelated band illustrates the general temperature and dopant dependences of ρ and S. As shown in Fig. 2, the high-temperature slope of the resistivity and the saturation limit of the thermopower decrease monotonically with increasing doping. such a dependence has been observed in $La_{2-x}M_xCuO_{4-y}$.[22,24] Alternate explanations for the linear resistivity are based on scattering by tunneling centers,[25] electron-electron scattering near a van Hove singularity,[26] and impurity scattering of Bose charge carriers.[27]

A strongly temperature-dependent specific heat coefficient $\gamma(T) = C(T)/T$ and a high-temperature Curie susceptibility also follow from the narrow band model.[20] Note that the zero-temperature mass enhancement m^*/m may be significantly larger than the value deduced from $\chi(T)$ or $\gamma(T)$ at high temperature.

Motivated by the experimental results cited above, we suggest a simple scenario for polaron formation. We assume that, at least in $La_{2-x}M_xCuO_{4-y}$, attention may be restricted to the two-dimensional CuO_2 layers. Mounting experimental evidence suggests that doping with divalent ions introduces holes in the oxygen $2p$ orbitals. A large intra-atomic Coulomb repulsion U_{dd} prevents double hole occupancy of the copper $3d$ orbitals. Hopping of $2p$ holes may proceed through direct overlap of near-neighbor oxygen orbitals, or by an indirect transfer through the copper sites. In either case, an O-O transfer matrix element $\tilde{t}$ of order 0.1 eV seems reasonable.

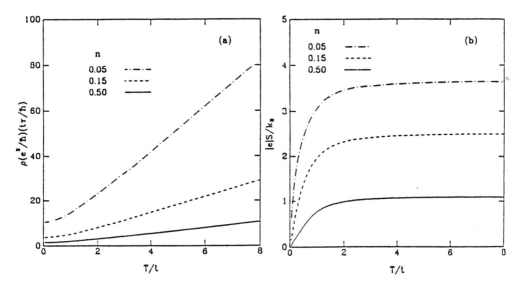

FIGURE 2 Resistivity (a) and thermopower (b) for a two-dimensional hole band with $\epsilon_k = -2t(\cos k_x + \cos k_y)$. A constant hole scattering rate τ^{-1} is assumed, and the chemical potential $\mu(T)$ is adjusted to keep n, the number of holes per site, fixed. Hole-hole correlations are ignored.

To discuss the coupling between $2p$ holes and lattice deformation in a simple approximation, we assume the heavy copper ions form a rigid lattice and that only the oxygen ions are mobile. If only a near-neighbor Cu-O spring constant is retained, there is, by symmetry, no tendency for bond relaxation when a $2p$ hole is introduced on an oxygen ion. For this reason, the near-neighbor O-O bond provides the dominant coupling to the lattice. The hole-phonon interaction takes the form

$$H_{\text{h-ph}} = \tilde{V} \sum_{\langle ij \rangle}(n_i + n_j)x_{ij} = V \sum_{\langle ij \rangle}(n_i + n_j)(a_{ij} + a_{ij}^+), \qquad (4)$$

where n_i counts the number of $2p$ holes on site i; x_{ij} is the O-O bond displacement; and a_{ij}^+ creates a bond phonon. Here, the coupling constant

$$V = \tilde{V}\sqrt{\frac{1}{2M\omega_0}} \qquad (5)$$

with ω_0 and M the phonon frequency and reduced mass. The complete Hamiltonian takes the form

$$H = H_{\text{ph}} + H_{\text{h-ph}} + H_{\text{h}}$$

$$H_{\text{ph}} = \omega_0 \sum_{\langle ij \rangle} a_{ij}^+ a_{ij} \qquad (6)$$

$$H_{\text{h}} = -\tilde{t} \sum_{\langle ij \rangle \sigma}(c_{i\sigma}^+ c_{j\sigma} + \text{h.c.}).$$

Solving $H_{\mathrm{ph}} + H_{\mathrm{h-ph}}$ for a given configuration of holes generates a pattern of static displacements in the oxygen lattice. The resulting static contribution to the polaron Hamiltonian is

$$
\begin{aligned}
H_{\mathrm{stat}} &= -\Delta \sum_{\langle ij \rangle} (n_i + n_j)^2 \\
&= -z\Delta \sum_i n_i - 2z\Delta \sum_i n_{i\uparrow} n_{i\downarrow} - 2\Delta \sum_{\langle ij \rangle} n_i n_j,
\end{aligned}
\tag{7a}
$$

where

$$
\Delta = \frac{V^2}{\omega_0} = \widetilde{V}^2 \cdot \frac{1}{2M\omega_0^2}
\tag{7b}
$$

and $z = 4$ is the oxygen-lattice coordination number. Since $(M\omega_0^2)^{1/2}$ is the static O-O spring constant, Δ is (at least nominally) independent of the oxygen ion mass. The last term in (7a) is a near-neighbor polaron attraction of the type first considered by Chakraverty.[3] The matrix element for polaron hopping between two lattice ground states is just

$$
t = \widetilde{t}\, e^{-(z-1)\Delta/\omega_0}.
\tag{8}
$$

Taking into account the Coulomb repulsion between holes, the full polaron Hamiltonian becomes

$$
\begin{aligned}
H_P &= -t \sum_{\langle ij \rangle \sigma} (c_{i\sigma}^+ c_{j\sigma} + \mathrm{h.c.}) + U_{\mathrm{eff}} \sum_i n_{i\uparrow} n_{i\downarrow} - V_{\mathrm{eff}} \sum_{\langle ij \rangle} n_i n_j \\
U_{\mathrm{eff}} &= U_c - 2z\Delta \\
V_{\mathrm{eff}} &= 2\Delta - (V_c + \delta V),
\end{aligned}
\tag{9}
$$

where U_c and V_c are on-site and near-neighbor Coulomb repulsions; and $\delta V (\propto t^2/\omega_0)$ is an additional repulsion which emerges in second-order perturbation theory.[28] For V_{eff} positive and larger than the polaron bandwidth, this Hamiltonian is expected to lead to bipolaron formation at high temperature and possible Bose condensation at low temperature.[29] For smaller V_{eff}, the model should instead lead to BCS-like superconductivity. The parameter regime for the high-T_c oxides may lie between these two extremes. Note that V_{eff} is an instantaneous coupling which incorporates Coulomb repulsion. As usual, the effect of U_c on superconductivity may be minimized by constructing a pseudopotential for the pair wave function.[30]

Oxygen isotope dependence enters through the hopping matrix element t (and through δV). Since the polaron bandwidth serves as the BCS energy cutoff, a detailed calculation is required to establish the isotope dependence of T_c for arbitrary band filling. (The conventional ladder-graph summation to find T_c is likely to be invalid, since no equivalent of Migdal's Theorem is available. We are currently carrying out Monte Carlo simulations for the model in equation (6).) The isotope dependence of the polaron bandwidth leads to interesting predictions for normal-state properties as well: decreasing the phonon frequency should increase the effective mass m^* and the high-temperature slope of the resistivity $d\rho/dT$.

Throughout this discussion, we have not referred explicitly to the role of the copper lattice. A more complete treatment, taking into account the hybridization of $Cu^{2+}O^{2-}$ and $Cu^{+}O^{-}$ states in the absence of doping,[31] seems essential for an explanation of magnetic properties. Our intent is simply to point out that polarons may play a role in determining the properties of the high-T$_c$ oxides.

ACKNOWLEDGMENTS

We would like to thank J. E. Hirsch for correcting some of our earlier ideas regarding the Holstein model and for discussing with us the results of Ref. 28. We would also like to acknowledge useful discussion with S. Sachdev and J. W. Wilkins. This work has been supported in part by the National Science Foundation under Grants DMR86–15454 and PHY82–17853 and by funds from NASA.

REFERENCES

1. J. G. Bednorz and K. A. Müller, *Z. Phys. B*, **64**, 18 (1986).
2. M. K. Wu, J. R. Ashburn, C. J. Torng, P. H. Hor, R. L. Meng, L. Gao Z. J. Huang, Y. Q. Wang, and C. W. Chu, *Phys. Rev. Lett.*, **58**, 908 (1987).
3. B. K. Chakraverty and C. Schlenker, *J. Phys. (Paris) Colloq.*, **37**, C4-353 (1976); B. K. Chakraverty, *J. Phys. (Paris) Lett.*, **40**, L-9 (1979).
4. K. Fossheim, T. Laegreid, E. Sandvold, F. Vassenden, K. A. Müller, and J. G. Bednorz, *Solid State Commun.* (1987), in press.
5. L. C. Bourne, A. Zettl, K. J. Chang, M. L. Cohen, A. M. Stacy, and W. K. Ham, *Phys. Rev. B*, **35**, (1987), in press.
6. D. McK. Paul, G. Balakrishnan, N. R. Bernhoeft, W. I. F. David, and W. T. A. Harrison, *Phys. Rev. Lett.*, **58**, 1976 (1987).
7. A. W. Sleight, J. L. Gillson, and P. E. Bierstedt, *Solid State Commun.*, **17**, 27 (1975).
8. E. Zirngiebl, J. O. Willis, J. D. Thompson, C. Y. Huang, J. L. Smith, P. H. Hor, R. L. Meng, C. W. Chu, and M. K. Wu, preprint (1987).
9. A. Junod, A. Bezinge, T. Graf, J.-L. Jorda, J. Muller, L. Antognazza, D. Cattani, J. Cors, M. Decroux, O. Fischer, M. Banovski, P. Genoud, L. Hoffmann, A. A. Manuel, M. Peter, E. Walker, M. Francois, K. Yvon, *Europhysics Lett.* (1987), in press.

10. A. Junod, A. Bezinge, D. Cattani, J. Cors, M. Decroux, O. Fischer, P. Genoud, L. Hoffmann, J.-L. Jorda, J. Muller, and E. Walker, *Proc. of LT-18* (1987), in press.

11. H. Maletta, R. L. Greene, T S. Plaskett, J. G. Bednorz, and K. A. Müller, *Proc. of LT-18* (1987), in press.

12. L. F. Mattheiss, *Phys. Rev. Lett.*, **58**, 1028 (1987).

13. J. Yu, A. J. Freeman and J.-H. Xu, *Phys. Rev. Lett.*, **58**, 1035 (1987); S. Massidda, J. Yu, A. J. Freeman, and D. Koelling (1987), preprint.

14. B. Batlogg, R. J. Cavas, A. Jayaraman, R. B. van Dover, G. A. Kourouklis, S. Sunshine, D. W. Murphy, L. W. Rupp, H. S. Chen, A. White, K. T. Short, A. M. Mujsce, and E. A. Rietman, *Phys. Rev. Lett.*, **58**, 2333 (1987).

15. L. C. Bourne, M. F. Crommie, A. Zettl, H.-C. zur Loye, S. W. Keller, K. L. Leary, A. M. Stacy, K. J. Chang, M. L. Cohen, and D. E. Morris, *Phys. Rev. Lett.*, **58**, 2337 (1987).

16. W. Weber, *Phys. Rev. Lett*, **58**, 1371 (1987).

17. J. Orenstein, private communication.

18. See, *e.g.*, I. G. Austin and N. F. Mott, *Adv. in Physics*, **18**, 41 (1969); or G. D. Mahan, *Many-Particle Physics*, Plenum, New York, 1981.

19. An alternate interpretation of the optical conductivity has been provided by C. M. Varma, S. Schmitt-Rink, and E. Abrahams, *Solid State Commun.* (1987) in press.

20. The magnetic susceptibility should exhibit Cure-law behavior ($\chi^{-1} \propto T$) in the same temperature range. A Curie law has been observed in some linebreak $YBa_2Cu_3O_{7-y}$ samples (see, *e.g.*, Refs. 9 and 21), but may be due to contamination with $BaCuO_2$.

21. R. J. Cava, B. Batlogg, R. B. van Cover, D. W. Murphy, S. Sunshine, T. Siegrist, J. P. Remeika, E. A. Rietman, S. Zahurak, and G. P. Espinosa, *Phys. Rev. Lett.*, **58**, 1676 (1987).

22. J. R. Cooper, B. Alavi, L.-W. Zhou, W. P. Behermann, and G. Grüner, *Phys. Rev. B*, **35** (1987), in press.

23. P. M. Chaikin and G. Beni, *Phys. Rev. B*, **13**, 647 (1976).

24. J. M. Tarascon, L. H. Greene, B. G. Bagley, W. R. McKinnon, P. Barboux, and G. W. Hull, *Proc. of the International Workshop on Novel Mechanisms of Superconductivity*, (1987).

25. E. Abrahams and C. M. Varma (1987), preprint.

26. P. A. Lee and N. Read, *Phys. Rev. Lett.*, **58**, 2691 (1987).

27. P. W. Anderson, B. Baskaran, Z. Zou, and T. Hsu, *Phys. Rev. Lett.*, **58**, 2790 (1987).

28. J. E. Hirsch and E. Fradkin, *Phys. Rev. B*, **27**, 4302 (1983).

29. A. Alexandrov and J. Ranninger, *Phys. Rev. B.*, **23**, 1796 (1981).

30. N. F. Berk and J. R. Schrieffer, *Phys. Rev. Lett.*, **17**, 433 (1966).

31. See *e.g.*, Ref. 19 and C. M. Varma, S. Schmitt-Rink, and E. Abrahams, *Proc. of the International Workshop on Novel Mechanisms of Superconductivity* (1987), and this volume.

Robert V. Kasowski, William Y. Hsu
Central Research and Development Department
E. I. du Pont de Nemours and Company
Experimental Station
Wilmington, Delaware 19898

and

Frank Herman
IBM Almaden Research Center
San Jose, California 95120-6099

Electronic Properties of Oxygen Vacancies in La₂CuO₄₋y[†]

The electronic properties of chemically pure La_2CuO_4 with and without oxygen vacancies have been investigated by the *ab initio* pseudofunction method using a supercell geometry. Oxygen vacancies alter the electronic structure significantly, breaking up the 2 eV-wide partially filled conduction band into narrower bands. For $La_2CuO_{3.75}$, the conduction bands near the Fermi level are only about 0.1 eV wide. Since the wide conduction band of stoichiometric La_2CuO_4 cannot support magnetic states according to accepted theories of magnetism, the present results suggest that antiferromagnetism observed in some non-stoichiometric La_2CuO_4 samples may be a consequence of the narrowing of bands produced by the presence of vacancies. Vacancy-induced narrow bands near the Fermi level may also play a role in raising T_c to 40 K via excitonic mechanisms. In chemically pure La_2CuO_4, the superconducting state is likely to occur between La_2CuO_4 and $La_2CuO_{3.75}$, the two limiting compositions studied here.

La_2CuO_4 exhibits a wide variety of interesting physical properties depending on how the samples are prepared[1-8]. La_2CuO_4 is metallic at room temperature and

[†]Reprinted from *Phys. Rev. B*, **36**, 7248 (1987).

above.[1,2] At low temperatures, La_2CuO_4 is variously found to be (a) metallic down to 4 K,[1] (b) semimetallic down to about 30 K, where the conduction electrons appear to become localized,[3] (c) semiconducting and weakly paramagnetic with the formation of a spin density wave around 250 K,[4] (d) magnetic, with a localized magnetic moment above 200 K which vanishes smoothly below 200 K,[5] (e) antiferromagnetic with a Néel temperature between 240 and 290 K,[6-8] and (f) superconducting below 40 K.[9,10]

Leaving aside deliberate doping, oxygen vacancies appear to be the most important variable controlling the final properties. For example, long-range antiferromagnetic order in oxygen-deficient La_2CuO_{4-y} is closely correlated to the value of y.[11] In this paper we will study the electronic properties of oxygen vacancies theoretically and discuss their possible relationship to T_c and antiferromagnetism in the La_2CuO_{4-y} system.

In recent theoretical studies,[12,13] we demonstrated that the orthorhombic phase of La_2CuO_4 is metallic, with a highly anisotropic band structure. The metallic character arises from the fact that the highest occupied and lowest unoccupied bands touch at certain points and along certain surfaces of the reduced zone. Elsewhere in the zone these bands are separated by a gap. In contrast to statements by earlier workers,[3,14,15] we suggested that the orthorhombic phase of La_2CuO_4 could be the host of superconductivity in doped as well as undoped samples under suitable conditions. These predictions have subsequently been confirmed experimentally,[9,10,16-19] as has our further prediction that at very low temperatures La_2CuO_4 becomes a semiconductor due to an orthorhombic to monoclinic structural phase transition.[20]

Since all band structure calculations to date indicate that La_2CuO_4 has broad bands crossing the Fermi level, the main paradox conceptually is the occurrence of localized moments and an anti-ferromagnetic ground state. The broad bands should preclude the existence of a magnetic ground state since the usual models of magnetism, such as the Hubbard and Anderson models, require narrow bands and large correlation energies.

In this paper we will show theoretically that oxygen vacancies alter the electronic structure significantly, breaking up and narrowing the the broad conduction bands of the stoichiometric compound in the neighborhood of the Fermi level. The vacancy-induced narrow band (in $La_2CuO_{3.75}$ is only 0.1 eV wide and so could conceivably support localized moments as well as an antiferromagnetic ground state. The narrow band obtained here is consistent with the earlier conclusion reached by one of us that O vacancies along one-dimensional Cu-O chains in $YBa_2Cu_3O_{7-x}$ can break the broad conduction bands at E_F into several narrow bands.[21] Such narrow bands are also of considerable interest because they are favorable for the excitonic superconductivity,[22] and so could contribute to the high T_c in these materials.

To study the effect of oxygen vacancies on the electronic properties of La_2CuO_4, we constructed a supercell composed of four tetragonal unit cells of La_2CuO_4, *i.e.*, 28-atoms. We used atomic coordinates appropriate for tetragonal La_2CuO_4 for ease of computation. The supercell was formed by doubling the tetragonal unit cell in the two planar directions. Body-centered tetragonal symmetry is preserved for this

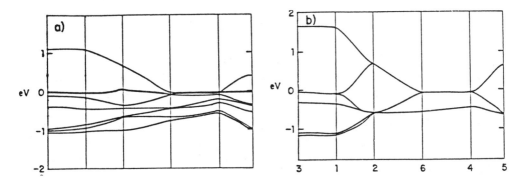

FIGURE 1 Band structures of La$_2$CuO$_{4-y}$: (a) y = 0.25 (with O vacancies) and (b) y = 0 (without O vacancies). The **k** points labelled 1 to 6 along the horizontal axis are: $3 = (.0, .0, .2878)$, $1 = (.0, .0, .0) = \Gamma$, $2 = (.25, .0, -.0179)$, $6 = (.25, .25, -.1438)$, $4 = (.25, .25, .2438)$ and $5 = (.0, .25, -.0179)$. Narrow band at $E_F (= 0$ eV) is highlighted.

configuration. We then removed 1 of the 16 O atoms in the supercell from the 2-D Cu-O network. The only remaining point symmetry is inversion. The separation between adjacent vacancies is 7.6 Å, and this proves to be sufficiently large to avoid significant intervacancy interactions.

We calculated the energy bands for this structure, formally La$_2$CuO$_{3.75}$, using the pseudofunction method.[23] Identical calculations were also carried out for the same structure but with the vacancy occupied. Although the actual symmetry is much higher now, we continued to use the lower symmetry of the 27-atom case as a numerical check. Our results for the 28-atom case (no vacancy) closely resemble published band structures for tetragonal La$_2$CuO$_4$,[14,15] thus increasing our confidence in the adequacy of our ordered vacancy calculation.

The two crystals we investigated, oxygen-deficient La$_2$CuO$_{3.75}$ and stoichiometric La$_2$CuO$_4$, bracket the undoped material La$_2$CuO$_{3.88}$, whose structure has been studied by neutron diffraction.[3] These studies clearly indicated that all the O vacancies occur in the Cu-O plane, as assumed in our theoretical model.

We used an ordered-vacancy supercell approach because we believe that this approach should provide accurate local information on the electronic structure of isolated oxygen vacancies. Moreover, none of presently available first-principles self-consistent band-structure computational methods can deal with disordered vacancies realistically. Since we found considerable changes in the electronic structure as the self-consistent iteration proceeded, we would expect simplified non-self-consistent methods such as empirical tight-binding calculations to miss many of the essential physical features of the (ordered or disordered) oxygen vacancy problem.

Our principal results are presented in Fig. 1. The bottom panel shows the energy bands for the 28-atom model (no vacancies), and the top panel the energy bands for the 27-atom (one O vacancy) model. It is readily apparent that, on the average, there is only one unfilled band for the 27-atom case and two for the 28-atom case. Apart from this feature, which is dictated by the different number of electrons and available orbitals, the band widths are strikingly different in the two cases. While there is a broad band roughly 2 eV wide crossing E_F for the 28-atom case, in agreement with previous studies of stoichiometric La_2CuO_4 by ourselves[12] and others,[14,15] the bands for the 27-atom case are considerably narrower, having been broken up by the ordered vacancy superlattice.

In contrast to textbook pictures of energy bands that are perturbed slightly only near the reduced zone boundaries by superlattice periodicity, the superlattice splitting extends throughout the reduced zone in the present instance. In fact, a band gap opens over a substantial part of the reduced zone between the 79th and 80th bands. Combined with the change in the number of filled bands (79 rather than 80), the oxygen vacancies create a very narrow band in the neighborhood of E_F whose width is about 0.1 eV. This narrow band forms the valence band edge and is highlighted in Fig. 1a.

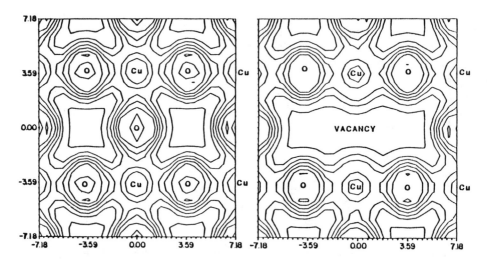

FIGURE 2 Pseudofunction charge density for La_2CuO_{4-y}: left panel is for $y = 0$ (without O vacancies) and right panel is for $y = .25$ (with O vacancies).

Charge density plots are useful for understanding the bonding qualitatively. In Fig. 2 we have plotted the total pseudofunction charge density for the 28-atom and 27-atom cases. We see that the charge density contours around the Cu and O atoms are quite similar, whether or not the O vacancy is present. The removal of an O atom thus has only a minor effect on neighboring atoms. Since there is very little rearrangement of electronic charge, it is as if a neutral O atom is plucked out of the lattice. The effect of removing the O atom is screened within a short distance, as is usually the case in covalent materials. Since the screening length is so short, adjacent vacancies do not interact with one another, suggesting that the present model describes the neighborhood of a vacancy realistically, assuming that lattice relaxation effects can be neglected.

In real oxygen-deficient compounds, the O vacancies are arranged at random, so that superlattice band gaps are not actually formed. Nevertheless, we would expect the electronic density of states in real materials to reflect the changes that occur in exaggerated form in our ordered vacancy model. That is to say, we would expect randomly arranged O vacancies to induce deep minima rather than absolute band gaps in the electronic density of states. So we will continue to speak of vacancy-induced narrow bands though we realize that these bands have smeared-out rather than sharply defined band edges.

In view of the narrow bands in our 27-atom model, and the fact that Cu 3d atomic orbitals contribute strongly to the crystal wave functions near E$_F$, we would expect this model to support Hubbard-type antiferromagnetism. We hope to carry out spin-polarized band calculations at a later date. If the above ideas are correct, we expect to find a non-magnetic ground state for the 28-atom structure and an antiferromagnetic ground state for the 27-atom structure.

Our conclusions should also apply to the lower symmetry orthorhombic phase because the crucial conduction bands under consideration are of the same width, about 2 eV, for both materials. In addition, spin-polarized band spitting, if it occurs in tetragonal La$_2$CuO$_{4-y}$, would occur more easily in the orthorhombic phase due to the preexistence of energy gaps in substantial portions of the reduced zone.

Emery[24] and Harrison[25] have concluded independently that pure La$_2$CuO$_4$ is an antiferromagnetic semiconductor at low temperature. Both use model Hamiltonians with electron-electron repulsion $U = 6.0$ eV. In their models a large U is necessary because the bands at E_F are very broad. Our calculations suggest that the presence of O vacancies leads to a reduction in the effective widths of energy bands near E_F, requiring smaller (and possibly more realistic) values of U for antiferromagnetic behavior. In any event, the pure stoichiometric compound should not be an antiferromagnetic semiconductor in view of the broad energy band.

The relationship between band narrowing and O vacancies should also play an important role for YBa$_2$Cu$_3$O$_{7-y}$. Depending on the conditions under which O is removed, YBa$_2$Cu$_3$O$_{7-y}$ ($0.5 < y < 1$) can be either a superconductor or a semiconductor. We believe this is related to whether O atoms are removed from the dimpled Cu-O planes or not. Semiconducting samples probably would have substantial O vacancies in those planes whereas superconducting samples would have less vacancies.

In summary, our calculations indicate that O vacancies tend to break up the broad conduction band in La$_2$CuO$_4$ into much narrower bands in the neighborhood of E_F, increasing the likelihood of antiferromagnetic behavior, and also possibly increasing the possibility of exciton-enhanced high temperature superconductivity. With vacancy-induced narrow bands, we would be dealing not with virtual excitations between the Cu-O bonding and antibounding orbitals[22], but rather with excitations between occupied and unoccupied Cu-O antibonding states split apart by the O vacancies. Physically, such excitations represent charge transfer from the

Cu-O band to the O vacancies, analogous to the situation we described earlier for $YBa_2Cu_3O_{7-y}$.[21]

ACKNOWLEDGEMENTS

We thank V. J. Emery, W. E. Harrison, W A. Little, T. N. Rhodin, and M. Schluter for stimulating discussions. The work at IBM was supported in part by the Office of Naval Research.

REFERENCES

1. R. Saez Puche, M. Norton and W. S. Glaunsinger, *Mat. Res. Bull.*, **17**, 1429 (1982).
2. P. Ganguly and C. N. R. Rao, *Mat. Res. Bull.*, **8**, 405 (1973).
3. J. D. Jorgensen, D. G. Hinks, D. W. Capone II, K. Zhang, H.-B. Schuttler, and M. B. Brodsky, *Phys. Rev. Lett.*, **58**, 1024 (1987).
4. S. Uchida, H. Takagi, H. Yanagisawa, K. Kishio, K. Kitazawa, K. Fueki, and S. Tanaka, *Japan. J. Appl. Phys.*, **26**, L445 (1987).
5. K. K. Singh, P. Ganguly and J. B. Goodenough, *J. Solid State Chem.*, **52**, 254 (1984).
6. Y. Yamaguchi, H. Yamauchi, M. Ohashi, H. Yamamoto, N. Shimoda, M. Kikuchi, and Y. Syono, *Japan. J. Appl. Phys.*, **26**, L447 (1987).
7. D. Vaknin, S. K. Sinha, D. E. Moncton, D. C. Johnston, J. M. Newsam, C. R. Safinya, and H. E. King, Jr., *Phys. Rev. Lett.*, **58**, 2802 (1987).
8. S. Mitsuda, G. Shirane, S. K. Sinha, D. C. Johnston, M. S. Alvarez, D. Vaknin, and D. E. Moncton, preprint.
9. P. M. Grant, S. S. P. Parkin, V. Y. Lee, E. M. Engler, M. L. Ramirez, J. E. Vazquez, G. Lim, R. D. Jacowitz, and R. L. Greene, *Phys. Rev. Lett.*, **58**, 2482 (1987).
10. J. Beille, R. Cabanel, C. Chaillout, B. Chevallier, G. Damazeau, F. Deslandes, J. Etourneau, P. Lejay, C. Michel, J. Provost, B. Raveau, A. Sulpice, J.-L. Tholence, and R. Tournier, submitted to *Phys. Mat. Conden.*
11. D. J. Johnston, J. P. Stokes, D. P. Goshorn, and J. T. Lewandowski, preprint.
12. R. V. Kasowski, W. Y. Hsu, and F. Herman, *Solid State Commun.*, in press; R. V. Kasowski, W. Y. Hsu, and F. Herman, *Proceedings of Symposium on High Temperature Superconductors*, ed. D. U. Gubser and M. Schluter, Materials Research Society, Pittsburgh, 1987, p. 41–44.
13. R. V. Kasowski and W. Y. Hsu, *Advanced Cermanic Materials: Superconducting Oxides*, in press.

14. L. F. Mattheiss, *Phys. Rev. Lett.*, **58**, 1028 (1987).
15. J. Yu, A. J. Freeman and J.-H. Xu, *Phys. Rev. Lett.*, **58**, 1035 (1987).
16. R. J. Cava, A. Santoro, D. W. Johnson, Jr., and W. W. Rhodes, *Phys. Rev. B*, **35**, 6717 (1987).
17. U. Geiser, M. A. Beno, A. J. Schultz, H. H. Wang, T. J. Allen, M. R. Monaghan, and J. M. Williams, *Phys. Rev. B*, **35**, 6721 (1987).
18. R. M. Fleming, B. Batlogg, R. J. Cava, and E. A. Rietman, *Phys. Rev. B*, **35**, 7191 (1987).
19. P. Day, M. Rosseinsky, K. Prassides, W. I. F. David, O. Moze, and A. Soper, preprint.
20. D. G. Gubser, *Proceedings of International Workshop on Novel Mechanisms in Superconductivity*, ed. V. Kresin, Plenum Press, New York, 1987, in press.
21. R. V. Kasowski, *Superlattices and Microstructures*, in press; W. Y. Hsu and R. V. Kasowski, *Proceedings of International Workshop on Novel Mechanisms in Superconductivity*, ed. V. Kresin, Plenum Press, New York, 1987, in press.
22. J. P. Coleman, J. T. McDevitt, and W. A. Little, preprint.
23. R. V. Kasowski, M.-H. Tsai, T. N. Rhodin, and D. D. Champliss, *Phys. Rev. B*, **34**, 2656 (1986).
24. V. J. Emery, *Phys. Rev. Lett.*, **58**, 2794 (1987).
25. W. A. Harrison, *Proceedings of International Workshop on Novel Mechanisms in Superconductivity*, ed. V. Kresin, Plenum Press, New York, 1987.

J. W. Halley
School of Physics and Astronomy
University of Minnesota
Minneapolis, Minnesota 55455

and

Herbert B. Shore
Department of Physics
San Diego State University
San Diego, California 92182

Model for the Role of Oxygen Defects in Oxygen Defect Superconductors[†]

We propose that oxygen defects provide a mechanism for pairing enhancement in oxide defect superconductors. A model is formulated and mean field Gorkov equations are derived and solved numerically for a two dimensional realization of the model. We show that very high temperature critical temperatures are not difficult to understand in such models. Several experimental features of the superconducting oxides, including spin glass-like properties, sensitivity to oxygen concentration and temperature dependence of positron annihilation times appear to be qualitatively consistent with such a model.

The recent discovery[1-2] of several high temperature oxides has led to two kinds of theoretical suggestions concerning the origin of the superconductivity. On one hand, several workers[3] are exploring whether superconductivity at such high temperatures can arise from electron-phonon interactions, as it does in normal superconductors or from the exchange of other bosonic excitations such as paramagnons, excitons or plasmons. On the other hand, Anderson[4] has proposed a resonating valence bond model in which the basic pairing interactions arise from Coulomb

[†]This is an expanded version of a paper appearing in *Phys. Rev. B*, **37**, 525 (1988).

interactions. Other workers[5] including Heisenberg[6] have proposed in a more general context that superconductivity can arise from pairing of electrons localized in a Wigner-like lattice through the action of direct Coulomb interactions between electrons without the mediation of phonons. In this note we suggest that, in the new oxide superconductors, oxygen vacancies are ideally configured to enhance interactions of pairs of electrons through partial bonding, in a way somewhat like the way that resonant bonds enhance pairing in the model of Ref. (4). In our model, however, we regard the local defect conformation as essential to achieve sufficient pairing. If this model is correct, the existence of isolated oxygen vacancies would be essential to the achievement of the observed high T_c. The oxygen vacancies in the $La_xSr_{1-x}CuO_{4-y}$ oxides are reported to strongly effect the superconducting properties.[7] Similar effects are reported[8] in $YBa_2Cu_3O_{9-x}$. Very recently,[9] there is evidence in positron annihilation studies that the charge density at oxygen vacancies in $YBa_2Cu_3O_{9-x}$ changes at the superconducting transition temperature.

To illustrate the possible effect of these vacancies on the superconductivity of the new oxides we show a plane of copper and oxygen sites in Fig. 1a. Planes of just this geometrical configuration exist in both classes of the high temperature superconducting oxides. The copper sites lie on the sites of a square lattice (circles) and the oxygen sites lie on the bonds (crosses). In Fig. 1a we also illustrate the geometry of a single isolated oxygen vacancy in such a plane (the site of the removed oxygen has been boxed and the two neighboring coppers have been shaded.) If the part of the crystal structure sketched in Fig. 1a was locally charge neutral before removal of an oxygen ion with a charge of approximately $-2|e|$, then the vacancy must attract electrons from elsewhere in the crystal in order to screen the charge imbalance created by the removal of the oxygen ion. Simple valence counting would suggest that the two (shaded) copper ions next to the vacancy had, before oxygen removal, empty d orbitals and that in zeroth order the two screening electrons which come in to screen the charge when a vacancy is created would fill two of these orbitals. (See Fig. 1b.) (This picture is recently confirmed by detailed band structure calculations which include regular arrays of oxygen vacancies.[10] Similar results would be expected when copper is replaced by other noble metals.)

The d-electrons of these two ions next to the vacancy will be strongly correlated because the outer electrons are weakly bound on each. The Heitler-London-like correlations arising from the close association which is forced on these electrons as they move in the relatively empty space between the two copper ions will be the fundamental origin of the pairing interaction which makes superconductivity possible in our picture as illustrated schematically in Fig. 1b. We suggest that the special structure of the oxygen defect together with the electronic filling of copper in particular combine to enhance the pairing interaction significantly in these materials. Thus our model can be said to propose a resolution of the dilema that the tendency of the coulomb interactions to induce pairing also tends to make the material a poor metal (or, in the worst case, an antiferromagnetic insulator): The ions next to vacancies induce pairing, but this pairing cannot lead to antiferromagnetism because the vacancies are disordered (or inappropriately ordered) and the material away

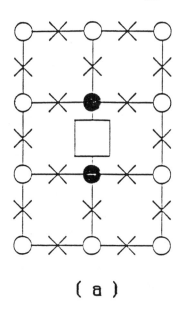
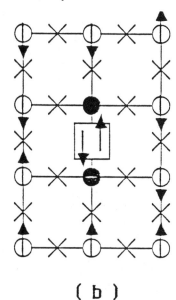

(a)   **(b)**

FIGURE 1 (a) Square lattice structure considered here. Open circles are copper ions. Crosses are oxygen ions. The two copper ions next to the vacancy are indicated by a dark circle. (b) Illustration of possible electronic structure of the vacancy. Pairs of electrons attracted to vacancy by coulomb interaction are strongly exchange correlated.

from the vacancies is metallic. Thus the pairing induced by sites next to vacancies leads to superconductivity.

In the compounds based on $LaCuO_4$ which show high temperature superconductivity, there is experimental evidence[11] that disorder on an extremely small length scale is significantly affecting the nature of the superconductivity. In the compounds of the type $YBa_2Cu_3O_{9-x}$ it is reported that, in addition to ordered oxygen vacancies between the planes, there are about 5% randomly distributed vacancies in the copper-oxygen planes.[12] Both the random vacancies in the planes and the ordered vacancies in the "chain" layers between planes could contribute to pairing by the mechanism considered here, though our detailed calculation will only apply to the former. In our model, we will assume that the vacancies are randomly distributed in the copper oxide planes and think of the mechanism by which the vacancies induce superconductivity as follows (see Fig. 2): one electron energies attract electrons to the pairs of copper ions next to vacancies. As described above, electrons at these sites have an enhanced probability to be found between the two ions with singlet spin correlation. In the case of an isolated vacancy, the resulting pair correlations can be regarded as extending some distance ξ away from the vacancy. If another vacancy, also inducing pair correlations, is found within a distance of order ξ, then the pair correlations can extend across the crystal in a way reminiscent of percolation and of theories of granular superconductors,[13] though we emphasize that here the disorder is microscopic in nature. Aspects of this picture have an obvious similarity to many reported experimental phenomena including

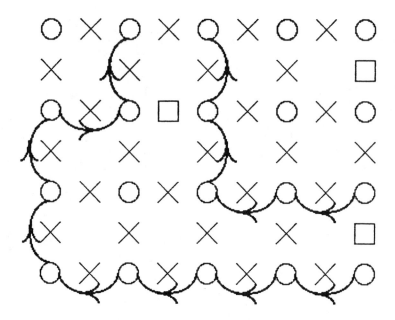

FIGURE 2 Schematic picture of the mechanism by which vacancies induce pairing in the model considered here.

spin glass-like behavior, dependence of the superconducting properties on the oxygen content of the oxide and recent reports of shortening of positron lifetimes at the superconducting transition. The picture suggests that an optimum vacancy concentration exists: adjacent vacancies will not enhance the pairing and the requirement that there not be too many adjacent vacancies will limit the vacancy concentration which will enhance T_c by this mechanism.

To formulate a model containing these ideas we work in a basis of tight binding states with one orbital state per "copper" site i. Creation operators for these orbitals are denoted $d_{i,\sigma}^\dagger$ where σ is a spin index. We think loosely of these orbitals as representing the last d orbital which remains unfilled in the Cu ion. We consider the Hamiltonian

$$H = \mu N = \sum_{i,\sigma} \epsilon(i) d_{i,\sigma}^\dagger d_{i,\sigma} + \sum_{i,j,\sigma} \left(b(i,j) d_{i\sigma}^\dagger d_{j,\sigma} + \text{h.c.} \right) + . \tag{1}$$

$$- \sum_{i,j} (J(i,j)/2) \left(d_{i\uparrow}^\dagger d_{j\downarrow}^\dagger - d_{i\downarrow}^\dagger d_{j\uparrow}^\dagger \right) (d_{j\downarrow} d_{i\uparrow} - d_{j\uparrow} d_{i\downarrow}) + \text{triplet terms}$$

The interaction terms have been written in a form which explicitly exhibits the energy advantage $J(i,j)$ which we assume accrues to a singlet pair at the site pair i, j relative to the average energy of a singlet pair on any site pair. A similar expression appears in studies of the resonating valence bond.[14] We will suppose, however, that the factor $J(i,j)$ is much larger when the sites i, j are nearest neighbors surrounding a "vacancy" where vacancies are distributed at random on the bonds of the square lattice and the site i, j sit on the sites of the same square lattice. The "triplet terms" will be dropped, assuming that the energy requirement for forming a triplet at any site pair is much larger than it is for forming a singlet. If we were to keep these terms under the assumptions made here, then they would oscillate rapidly in time in the calculation reported below and would not contribute to the result. Put another way, we confine attention to the lowest energy subspace in which triplet correlations of the spins of electrons on sites on each side of a vacancy do not occur. For sites which are not adjacent to oxygen vacancies, we have, in the model calculations presented here, dropped the two electron term J. This last assumption is qualitatively correct if, as we assume, the exchange coupling is larger nearer vacancy sites than elsewhere, but it oversimplifies the description of the electron dynamics away form the vacancy sites. A negative pairing energy of the sort described by the J term might also arise from local electron phonon interactions as in "negative U" theories, though we regard this as less likely than exchange as the origin of J in the superconducting oxides. In addition to the exchange contribution to the J term (which we have argued should lead to a positive J corresponding to an attractive interaction) there will also be a contribution associated with the direct coulomb integral (the off site Hubbard U). This contribution will reduce J or make J negative, that is repulsive. It seems possible to us that even in this case, our model may lead to a BCS like state through a mechanism like that described in Ref. 16. We have not yet found such a solution, however, and in the rest of this paper we will assume that J is positive.

We suppose that the order parameter of interest is the expectation value $\langle (d_{j\downarrow}d_{i\uparrow} - d_{j\uparrow}d_{i\downarrow}) \rangle = \Delta(i,j)$. A similar assumption is made in Ref. 14 but here we regard the oxygen vacancies as stabilizing these quantities. With these assumptions we construct a mean field theory for the superconducting state which is of the BCS form. The main distinctions between the resulting formulation and more conventional ones are that the pairing function is local and the theory is formulated in position space, instead of momentum space. Our formulation shares these distinctions with Ref. 4 and 5.

For simplicity we make the following additional assumptions: We suppose that $J(ij)$ is small away from the vacancy pairs and set it equal to zero there. At vacancy sites the value of $J(\text{ij}) = J$ is a parameter of the model. We keep only nearest neighbor terms everywhere. b_{ij} is made the same at all nearest neighbor pairs and set equal to b, a second parameter of the model. $\epsilon(i)$ is zero at sites away from vacancies (half filled band) but negative (note that ϵ includes the term $-\mu$) on sites next to oxygen vacancies. The value of ϵ at copper sites next to oxygen sites is the third parameter of the mode. We define the retarded functions:

$$G_\sigma^{ij}(t) = -i\Theta(t)\langle \{d_{i,\sigma}(t), d_{j,\sigma}^\dagger(0)\} \rangle$$
$$F_\sigma^{ij}(t) = -i\Theta(t)\langle \{d_{i,\sigma}^\dagger(t), d_{j,-\sigma}^\dagger(0)\} \rangle$$

We obtain equations of motion for the functions $G_{ij} = G_{\uparrow}^{ij} + G_{\downarrow}^{ij}$ and $F_{ij}(t) = F_{\uparrow}^{ij} - F_{\downarrow}^{ij}$:

$$i\hbar d\,G_{ij}/dt$$
$$= 2\hbar\delta(t)\delta_{i,j} + \epsilon(i)G_{ij} + \sum_{\delta} b(i+\delta,i)G_{i+\delta j} + \sum_{\delta}(J(i+\delta,i)/2)\Delta_{i,\delta}F_{i+\delta j}. \quad (2)$$

$$i\hbar d F_{ij}/dt$$
$$= -\epsilon(i)F_{ij} - \sum_{\delta} b(i+\delta,i)F_{i+\delta j} + \sum_{\delta}(J(i+\delta,i)/2)\, Delta_{i,\delta}^{*}G_{i+\delta j}. \quad (3)$$

with the initial conditions $G_{ij}(0) = -i2\delta_{ij}$, $F_{ij}(0) = 0$. The gap equation is

$$\Delta_{ij}^{*} = -i\int_{-\infty}^{\infty} \frac{F_{ij}(\omega+i\epsilon) - F_{ij}(\omega-i\epsilon)}{e^{\beta\omega}+1}d\omega. \quad (4)$$

where

$$F_{ij}(\omega+i\epsilon) = 1/2\pi \int_{-\infty}^{\infty} F_{ij}(t)e^{i\omega t}\,dt$$

and the analytical continuation defines $F_{ij}(\omega-i\epsilon)$ in the usual way.[17] To zeroth order (but not to first order) we will assume in the equation of motion that the magnitude of the gap is a fixed constant on the vacancy sites and zero otherwise. This zeroth order assumption does not mean that the order parameter is zero away from the vacancies in our calculations, because the equations of motion above which we solve imply that the function F becomes nonzero away from vacancies if Δ is finite at vacancies. This is a manifestation of the effect of propagation of the pairing order from one vacancy to the next which was discussed physically above. Numerical studies of the model when only two vacancy sites are present reveal that the model favors equal phases of the gap for vacancies on parallel bonds but phases of the gap differing by π for vacancies on perpendicular bonds. These studies suggested that, in zero order, we choose the gap to be real and of one sign on the vacancies which are on bonds in the x direction and real and of the opposite sign on vacancies which are on bonds in the y direction (see Fig. 1). The calculations described below confirmed that this choice gives a more stable superconducting state than other choices of the phases of Δ in zeroth order. We could lift the assumption of a constant input magnitude of Δ in more detailed calculations in this model without too much difficulty. In fact, Eq. (4) in our model is an integral equation, conveniently formulated in real space and Δ is a strong function of position in the lattice. This feature may prove useful in interpreting reports of measured superconducting gaps which depend strongly on the technique used to measure them. Microscopic calculations could be used to roughly estimate the remaining parameters ϵ, b and J. Our main point here is that ϵ, b and J are all electronic scale energies of the order of electron volts. ϵ is related to the position of the Fermi energy relative to the one electron energies in the vacancy sites.

We solve these equations using the equation of motion method.[18] In order to reduce the number of equations, we define the sums $F_i = \sum_j c_j F_{ij}$, $G_i = \sum_j c_j G_{ij}$. Then $F_i(0) = 0$, $G_i(0) = -i2c_i$. To calculate[18] the average Δ at the vacancy sites we choose the c_i as follows. All c_i associated with sites not next to a vacancy are zero. Associate a random number ϕ_i evenly distributed between 0 and 2π with each site next to a vacancy. Now consider a pair of "copper" sites next to a vacancy. Label the sites 1 and 2. Set $c_1 = e^{i\phi_2}$ and $c_2 = e^{i\phi_1}$. We stress that the phase factors $e^{i\phi}$ are a calculational device only and are not to be confused with the phase factors associated with the gap function itself. Choosing c_i equal to zero away from vacancies is done only because we are choosing to evaluate Δ self-consistently at those sites and does not mean that the gap is zero or neglected elsewhere on the lattice in our calculation. This approach is taken directly from Ref. 18.

With choices of the coefficients c_i described above, cancellation of random phases[18] gives the gap equation: We define $F_r(t) = \sum_i \pm e^{-i\phi_i} F_i$ where the plus sign is used if i is adjacent to an x bond and the minus sign is used if i is adjacent to a y bond. Then Eq. (4) takes the form

$$\Delta^* = \frac{-1}{2N_s \beta\hbar} \int_0^\infty \frac{F_r(t)}{\sinh(t\pi/\beta\hbar)} dt. \tag{5}$$

where N_s is the number of oxygen vacancies. By use of Eq. (2) and (3) we obtain the equations of motion for F_i and G_i which are essentially identical in form to Eq. (2) and (3).

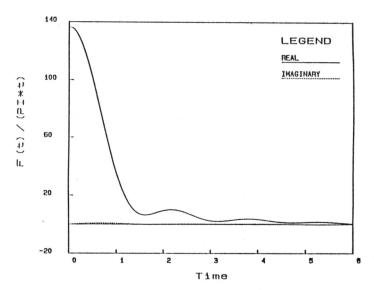

FIGURE 3 Negative of calculated integrand of Eq. (5) at zero temperature. 5% vacancies, $b = 1.0$, $\epsilon = 0.5$ $J = 1.75$. Solution for Δ with these parameters is $\Delta = 0.2423$.

We solve these equations by simple numerical integration in the time domain for a realization of the model on a 50×50 square lattice of the type shown in Fig. 1. In all calculations discussed below we took $\epsilon = 0.5$ and $b = 1.0$. We first studied the problem of one vacancy on this lattice numerically within the model. We find that for $J > 1.82$, the single vacancy problem has a solution with nonzero Δ at zero temperature and Δ goes to zero with increasing temperature in the standard BCS way. This result is an artifact of the mean field theory, since fluctuations would obviously destroy such an order parameter associated with just one site. We next studied two vacancy sites within the model numerically. As described above, these studies established that the phases of the gaps tend to equal signs when the bonds on which the vacancies sit are parallel and to opposite signs otherwise. These results for the single and pair vacancy problem establish that a meaningful test of whether the model leads to a phase transition of the sort envisioned qualitatively above requires that a self consistent solution to the Eq. (5) exist with $J < 1.82$ when $\epsilon = 0.5$ and $b = 1.0$. With 5% vacancies on the 50×50 lattice we find such solutions at least for $J > 1.6$, thus establishing that the model does lead to superconductivity of the type envisioned for appropriate parameters. We show the integrand $\mathbf{F_r}/\pi t$ of the right hand side of Eq. (5) in Fig. 3 for such a self-consistent solution when $J = 1.75$. For this value of J, the long time tail of the function $\mathbf{F_r}/\pi t$ is providing the needed weight to make the right hand side of Eq. (5) large enough to achieve self-consistency, whereas for $J > 1.82$, a long tail (arising from interactions between vacancies) is not required to achieve self-consistency. For $J = 1.75$, calculation of the temperature dependence of Δ gives a T_c of 0.152. Since b is of order one eV this would be a very large T_c, but the value of J or the value of p may be unrealistically large. Reducing either J or p could reduce T_c to a level consistent with experimental observations. Such fitting would not be warranted at present without more detailed experimental or theoretical information on these quantities. To find solutions for much smaller values of J is numerically more difficult (because the integrand of Eq. (5) develops a long tail) but we have preliminary indications that solutions exist down to very small values of J. For the value $J = 1.75$ we have also studied the dependence of the solutions to the gap equation on the vacancy concentration p. As p decreases, the zero temperature gap and the transition temperature decrease and become zero at about $p = 2\%$.

We present a plausibility argument to show how our model might account for the temperature dependence of the positron annihilation rate reported in Ref. 9 (see Fig. 4). We suppose, as in the present model that the result of the random distribution of vacancies is a peak in the density of states of width Γ centered an energy $\bar{\epsilon}$ from the fermi surface. If, as we assume , the superconductivity is a consequence of the pairing of the electrons, largely localised near the oxygen vacancies, which lie in this peak, then one has from a mean field BCS model that the number $n_{sc}(\vec{r} = 0)$ of electrons at a vacancy is very approximately

$$n_{sc}(\vec{r} = 0) \approx \frac{1}{2} \int \left[1 - \frac{\omega}{\sqrt{\omega^2 + \Delta^2}} \tanh\left(\frac{\sqrt{\omega^2 + \Delta^2}}{2T} \right) \right] \frac{\Gamma d\omega}{(\omega \bar{\epsilon})^2 + \Gamma^2}$$

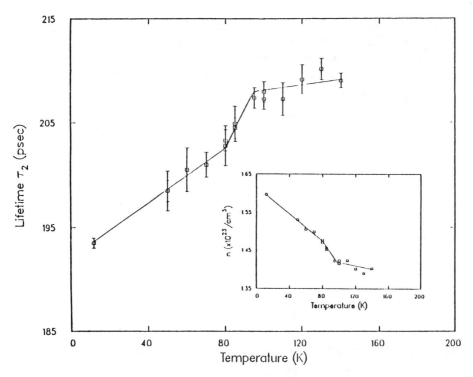

FIGURE 4 Data on positron annihilation rate in $YBa_2Cu_3O_{7-x}$. The inset shows the deduced temperature dependence of the electron density at vacancy sites. (Ref. 9)

Thus, if $\bar{\varepsilon}, \Delta \ll T$, the change at T_c is

$$n_{sc}(\vec{r}=0) - n_N(\vec{r}=0) = \frac{\Delta^2\bar{\varepsilon}}{48T^3} + \cdots \propto T_c - T$$

as long as $\Delta, \varepsilon \ll T$. A more careful calculation of this quantity using our model is clearly necessary.

Among the many problems which a full solution of such a model raises, we note that we have not proved in detail that the equations lead to a Meissner effect or to zero resistance below T_c. With regard to the Meissner effect we note that the electromagnetic vector potential will enter the equations in the same way in which it does in the Gorkov formulation of BCS theory and therefore a Meissner effect is to be expected. Demonstration of zero resistance is more difficult (as indeed it was for the BCS theory) but we comment that as long as the energy level spectrum has a gap tied to the Fermi surface and the normal system (before the transition) is metallic, a zero resistance state is to be expected. Here, both these criteria are satisfied. (The gap in the energy level spectrum will not be the quantity $J\Delta$ which we calculated self-consistently above, but rather will be related to the smallest value which the function F takes between the vacancies.) We note that if, as expected, the zero temperature coherence lengths are short, then the mean field treatment outlined here will not be adequate for a detailed description of the thermodynamics. At the mean field level, a full solution will lead to amplitude and phase variations of the gap with position which can be expected to give rise to spin-glass like behavior[11] arising from the randomness of the vacancies on an atomic scale.

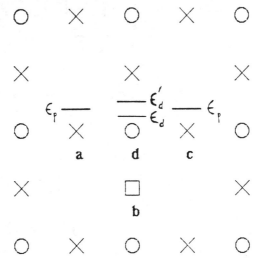

FIGURE 5 Possible structure of one electron site energies of holes near an oxgen vacancy.

Finally, we add a speculative remark concerning the effects of oxygen vacancies in the case that the current carriers are holes on the oxygen ions (as suggested by the models of Section IV of this volume for example). In the Appendix we describe a mechanism by which the exchange between holes on oxygen ions would be enhanced by the presence of a nearby oxygen vacancy. If this effect of exchange enhancement is large then the form of a mean field theory taking it into account would be qualitatively very similar to the model we have reported above.

ACKNOWLEDGEMENTS

We are grateful to R. Schrieffer, P.-G. de Gennes, R. Palmer, A. Goldman, F. Herman, W. Hsu, K. G. Lynn, M. Strongin, S. Oseroff and N. E. Bickers for discussions. This work was performed while J. W. H. was recipient of the Paul Flory sabbatical award at the IBM Almaden Research Center in San Jose.

APPENDIX

In Fig. 5a we sketch the environment of a copper ion and in Fig. 5b we show the same region of the plane when one of the oxygen ions is replaced by an oxygen vacancy. The energies of holes on the sites occupied by oxygen ions are indicated by ϵ_p while the energy of a hole on the copper ion is denoted ϵ_d. In the absence of a vacancy and when the d band is half filled, then the copper ions are in the Cu^{2+} state and the d hole level is occupied by one hole and lies below the fermi level. When there is a vacancy adjacent to the copper ion, then, as we argued above, the copper ion attains a valence of $+1$ because of the screening of the vacancy and the d-hole level on the copper is pushed up and becomes empty. In the models of Section IV, holes are introduced onto the oxgen sites by doping away from 1/2 filling. The attractive interaction arising from exchange between these holes is estimated to have a contribution (see Eq. (4) of Ref. 18) of

$$v_O \sim \frac{b^4}{\left(U_d - (\epsilon_p - \epsilon_d)\right)^2} \frac{1}{J_{\text{Cu-Cu}}}$$

which is said to give rise to superconductivity. Now consider the effect of a vacancy (say at site b as shown in Fig. 5b). In accordance with the arguments of this paper and the band structure calculations of Ref. 10, electrons are attracted to the adjacent Cu^{2+} ions making them Cu^{1+}-like. In the hole language the hole is unoccupied or partially occupied. The interchange process leading to the attraction between holes on oxygen sites a and c of Fig. 5 as described in Ref. 18 and leading to the preceding equation now involves an intermediate state with one hole (not two) on the copper site d. Thus U_d disappears from the expression for the attraction, which becomes $v'_O \sim \frac{b^4}{(\epsilon_p - \epsilon'_d)^2} \frac{1}{J_{\text{Cu-Cu}}}$. Because U_d is the largest energy in the problem ($\sim 8\,\text{eV}$), this effect could greatly enhance v'_O.

REFERENCES

1. J. G. Bednorz and K. A. Muller, *Z. Phys. B*, **64**, 189 (1986).
2. M. K. Wu, J. R. Ashburn, C. J. Torng, P. H. Hor, R. L. Meng, L. Gao, Z. J. Huang, Y. Q. Wang, and C. W. Chu, *Phys. Rev. Lett.*, **58**, 908 (1987).
3. H.-B. Schuttler, J. D. Jorgenson, D. G. Hinks, D. W. Capone, and D. J. Scalapino, *Phys. Rev. Lett.*, **58**, 1147 (1987).
4. P. W. Anderson, *Science*, **235**, 1196 (1987).
5. L. G. J. Van Dijk, and G. Vertogen, *Physics Letters*, **A115**, 63 (1986).
6. W. Heisenberg, *Z. Naturforsch*, **2a**, 185 (1947).
7. J. M. Tarascon, L. H. Green, W. R. Greene, G. W. Hull, T. H. Geballe, *Science*, **235**, 1373 (1987).

8. P. K. Gallagher, H. M. O Brien, S. A. Sunshine, D. W. Murphy, *Mat. Res. Bull.*, in press.

9. K. G. Lynn, preprint.

10. R. V. Kasowski, W. Y. Hsu, and F. Herman, *Phys. Rev. B*, **36**, 7248 (1987), and this volume.

11. K. A. Muller, M. Takashige, and J. G. Bednorz, *Phys. Rev. Lett.*, **58**, 1143 (1987); S. Oseroff, private communication.

12. I. Schuller, private communication.

13. C. Ebner and D. Stroud, *Phys. Rev. B.*, **31**, 165 (1985).

14. G. Baskaran, A. Zou, and P. W. Anderson, Solid State Commun.**63**, 973 (1987) and this volume. 63

15. W. Kohn and J. M. Luttinger, *Phys. Rev. Lett.*, **15**, 524 (1965).

16. D. N. Zubarev, *Sov. Phys. Uspekhi*, **3**, 320 (1960).

17. R. Alben, M. Blume, H. Krakauer, and L. Schwartz, *Phys. Rev. B*, **12**, 4090 (1975).

18. V. J. Emery, *Phys. Rev. Lett.*, **58**, 2794 (1987) and this volume.

RESONATING VALENCE

BOND MODELS

G. Baskaran, Z. Zou and P. W. Anderson
Joseph Henry Laboratories of Physics
Jadwin Hall, Princeton University
Princeton, New Jersy 08544

The Resonating Valence Bond State and High-T_c Superconductivity —A Mean Field Theory[†]

A mean field type theory is developed for the insulating RVB state and high temperature superconductivity in doped La_2CuO_4 and other high-T_c oxides. The T_c as a function of the doping parameter δ drops sharply around $\frac{t}{U} \sim \delta$. The zero temperature gap versus T_c relation depends sensitively on the choice of the parameters. Gutzwiller projection, supression of T_c by phase fluctuations for small δ and screening effects are briefly discussed.

The recent discovery[1] of high-T_c superconductivity in a number of doped lanthanum copper oxides has generated strong interest in the basic properties of these substances and possible new mechanisms[2,3] for superconductivity. High temperature susceptibility data[4] on insulating La_2CuO_4 suggests that Cu^{2+} is in an $S = 1/2$, orbitally non-degenerate state, with a copper $d_{x^2-y^2}$ orbital strongly hybridizing with the neighboring oxygen p-levels. Based on the magnetic data of Ganguly and Rao, a predominately electronic and magnetic mechanism for superconductivity in these compounds has been suggested by Anderson.[2] It is hypothesized that the insulating state of pure La_2CuO_4 is the resonating valence bond (RVB) state or quantum spin liquid proposed in 1973.[5] There are pre-existing spin

[†]Reprinted from *Solid State Commun.*, **63**, 973 (1987).

singlet pairs in the RVB state and they become charged superconducting Cooper pairs by strong enough doping.

In this communication we present a mean field theory for the RVB state and the high-T_c superconductivity. Our theory explicitly takes into account the singlet correlations induced between electrons on neighboring sites by the Hubbard U and shows how superconductivity results in a model with pure repulsive interaction. It is shown that the general picture seems to be consistent with what was proposed by Anderson.[2] We found that the RVB wave function for the insulator comes out as a self-consistent solution to the problem. In the insulating phase there exist neutral fermionic excitations with effective mass of the order of J^{-1}, where J is the antiferromagnetic exchange integral. Charge excitations are projected out by the Gutzwiller procedure, so that the state behaves like a Fermi liquid with $F_0^s = \infty$. We calculate T_c as a function of doping parameter δ; it has sharp fall around $t/U \sim \delta$, where t and U are the usual "Hubbard" parameters. The relation between the zero-temperature gap and T_c is found to be very sensitive to the parameters. At the end we discuss the Gutzwiller projection, and how much it may change the picture. We also discuss phase fluctuations and suppression of T_c for small δ. The screening of the Hubbard U is also briefly discussed. For the exact half filled band case our theory has some mathematical resemblence to a theory recently discussed by Noga.[6] We also would like to point out that antiferromagnetic spin-fluctuation-mediated even-parity pairing in heavy fermion superconductors has been discussed by Miyake et al.[7]

Since the high-T_c superconducting transition appears near a metal-insulator transition, it is believed that a nearly half-filled "Hubbard" model describes the system:

$$H = -t \sum_{\langle ij \rangle}(C_{i\sigma}^+ C_{j\sigma} + \text{h.c.}) + U \sum_i n_{i\uparrow} n_{i\downarrow} \tag{1}$$

and we will explicitly introduce a chemical potential μ modeling the doping process later. We consider a 2-dimensional square lattice and perform a mean field theory. The usual argument of weaker coupling between Cu Layers can be used to stabilize the 2-d mean field phase that we find. After a canonical transformation,[8,9] we obtain an approximate effective Hamiltonian which is defined only in the non-doubly occupied subspace:

$$H = -t \sum_{\langle ij \rangle}(1 - n_{i-\sigma})C_{i\sigma}^+ C_{j\sigma}(1 - n_{j-\sigma})$$
$$+ \mu \sum_i C_{i\sigma}^+ C_{i\sigma} + J \sum_{\langle ij \rangle}(\vec{S}_i \cdot \vec{S}_j - 1/4 n_i n_j) \tag{2}$$

with $J = 4t^2/U$, $S_i^+ = C_{i\uparrow}^+ C_{i\downarrow}$, etc. For a half-filled band in the singly occupied site subspace, Eq. (2) becomes $H = J \sum_{\langle ij \rangle}(\vec{S}_i \cdot \vec{S}_j - 1/4)$. We will concentrate on the Hamiltonian (2) which can be used for the non-half filled case. We will relax the restriction to singly occupied subspace and work in the entire Hilbert space for pure mathematical convenience and to get a general idea of the physics of the situation. Part of this restriction is taken into account by approximating the hopping term $-t(1-n_{i-\sigma})C_{i\sigma}^{+}C_{j\sigma}(1-n_{j-\sigma})$ by $-t\delta C_{i\sigma}^{+}C_{j\sigma}$, where δ is the fractional difference of n from the half-filled case. We will discuss at the end, using a Gutzwiller projection, the effect of removing the double occupancy restriction. Thus the Hamiltonian can be written in terms of fermion operators:

$$
\begin{aligned}
H = &- t \sum_{\langle ij \rangle}(C_{i\sigma}^{+}C_{j\sigma} + \text{h.c.}) \\
&+ J \sum_{\langle ij \rangle}\left[\frac{1}{2}(C_{i\uparrow}^{+}C_{i\downarrow}C_{j\downarrow}^{+}C_{j\uparrow} + \text{h.c.}) \right. \\
&\left. + \frac{1}{2}(n_{i\uparrow}n_{j\downarrow} + n_{i\downarrow}n_{j\uparrow})\right]
\end{aligned} \tag{3}
$$

We note that the two-body term can be written in terms of valence bond "singlet" pair creation and annihilation operators defined by

$$
b_{ij}^{+} = \frac{1}{\sqrt{2}}(C_{i\uparrow}^{+}C_{j\downarrow}^{+} - C_{i\downarrow}^{+}C_{j\uparrow}^{+}), \tag{4}
$$

$$
H = - t\delta \sum_{\langle ij \rangle}(C_{i\sigma}^{+}C_{j\sigma} + \text{h.c.}) - J \sum_{\langle ij \rangle}b_{ij}^{+}b_{ij} \tag{5}
$$

where $n_i = n_{i\uparrow} + n_{i\downarrow}$. The negative sign of the second term above suggests that the singlet objects (approximate bosons) will try to undergo "Bose condensation" into zero momentum state. The numerical simulations of Hirsch[8] strongly suggests that the bond singlet pairing tendency dominates. Hirsch's results also suggests that when the Hubbard U is not very large, the RVB state may be stabilized by quantum fluctuations for the exactly half filled band.

When we go to momentum space and make a Hartree-Fock factorization in Eq. (5), the resulting Hamiltonian is

$$
\begin{aligned}
H = &\sum_{\vec{k}\sigma}(\epsilon_k - \mu)C_{k\sigma}^{+}C_{k\sigma} \\
&- J \sum_{k}(\Delta\gamma_k C_{k\uparrow}^{+}C_{-k\downarrow}^{+} + \text{h.c.}) \\
&+ N(\Delta^2 + p^2)
\end{aligned} \tag{6}
$$

where the self-consistent order parameters Δ and p are defined as $\sqrt{2}\langle b_{ij}\rangle = \Delta$, $p = \langle C_{i\sigma}^{+}C_{j\sigma}\rangle$, for $\langle ij \rangle$ nearest neighbors and zero otherwise; $\epsilon_k = -(2t\delta + pJ)(\cos k_x a + \cos k_y a)$ for two dimensions ; $\gamma_k = -\epsilon_k/(2t\delta + pJ)$. The Hamiltonian is of BCS type. Diagonalizing Eq. (6) by Bogoliubov transformation gives quasiparticle energy

$$
E_k = \sqrt{(\epsilon_k - \bar{\mu})^2 + J^2\gamma_k^2\Delta^2}. \tag{7}
$$

The corresponding quasiparticle operators, α_k, β_k are related to $C's$ by a unitary transformation , and therefore they are fermions. The Hartree-Fock order parameters δ and p are to be found self-consistently by minimizing the free energy. The gap and chemical potential equations are:

$$\frac{1}{N} \sum_k \frac{\tanh \beta E_k/2}{E_k} \gamma_k^2 = \frac{2}{J} \tag{8}$$

$$\frac{1}{N} \sum_k \frac{\tanh \beta E_k/2}{E_k} (\epsilon_k - \mu) = \delta \tag{9}$$

$$p = -\frac{1}{2N} \sum_k \frac{\tanh \beta E_k/2}{2E_k} \gamma_k (\epsilon_k - \mu)$$

Before we discuss the general solution to Eq. (8, 9), let's first look at some of the unusual consequence of this simple theory in the insulating phase in which $\delta = \mu = 0$. It's easy to see that Eq. (9) is trivially satisfied with $p = 0$ and the quasiparticle operators are given by

$$\alpha_k = \begin{cases} \frac{1}{\sqrt{2}} C_{k\uparrow}^+ + C_{-k\downarrow}^+, & \text{if } |k| < k_F \\ \frac{1}{\sqrt{2}} C_{k\uparrow}^+ - C_{-k\downarrow}^+, & \text{if } |k| > k_F \end{cases} \tag{10}$$

Similar relations hold for β_k. The quasiparticle energy becomes

$$E_k = \Delta J |\gamma_k| = \Delta J |\cos k_x a + \cos k_y a| \tag{11}$$

Thus we get from first principles the RVB state constructed by Anderson

$$|G\rangle = P_N \prod_{\vec{k}} (u_k + v_k C_{k\uparrow}^+ C_{-k\downarrow}^+ |0\rangle$$

with

$$\sum_k \frac{v_k}{u_k} = \sum_k a(k) = 0; \qquad |a(k)| = 1 \tag{12}$$

Note that $a(k)$ does change sign across a Fermi surface as is pointed out by Anderson.

As a consequence of the signs in Eq. (10), $\alpha_k^+ \alpha_{k'}^+$, creates a charged excitation if $\vec{k}$ and $\vec{k}'$ are both on the same side of the pseudo Fermi surface, and a neutral spin excitation if they are on opposite sides. When we project out charged excitations in the insulating state, only the latter spectrum will persist; it will resemble closely the excitation spectrum of a simple Fermi gas. There will be a gap for charged excitations, which will more or less gradually disappear with temperature. Spin excitations will have no gap, but their spectrum is restricted in $\vec{k}, \omega$ space by the condition that $\vec{k}'$ and $\vec{k}$ be on opposite sides of the pseudo Fermi surface.

The excitation spectrum (11) implies a linear temperature dependence of low temperature specific heat due to the fermionic nature of the quasiparticles, $C(T) \propto$

γT, where γ is proportional to the effective mass of the quasiparticle which is of order of J^{-1}. In the presence of a weak magnetic field the degeneracy of the two branches will be lifted giving rise to a Pauli-like susceptibility. Thus the experimental measurement of specific heat and susceptibility of the pure La_2CuO_4 at low temperature would be a decisive test of the present theory. When $T > 3J/16(1/N\sum_k \gamma_k^2)$, Δ vanishes, and the susceptibility obeys a Curie-Weiss law, since an energy J is needed to break up a valence bonded pair. This has been partially confirmed by experiment (see Ref. 4).

It is important to recognize that there can be no true "ODLRO" for exactly half-filled band and that the phase of the mean field Δ is to be averaged over. Thus there can be no BCS type transition in the insulating case. In essence, the phase correlation length for charged excitations is zero, and as we remark later the corresponding reduction in T_c when charge excitations are present is striking (as well as the low critical currents that it implies). We see no way of formally expresing the fact that spin excitations nonetheless exhibit a sharp Fermi surface except to appeal to the theoretical possibility that $F_0^s \to \infty$.

Now let's look at the superconducting state in which the band is non-half filled due to doping. The quasiparticle energy is given by Eq. (11). From the mean field point of view, as soon as the system is metallized, it should become superconducting and open up a gap. However, in the low doping concentration limit ($\delta \ll 1$), the phase fluctuations play an important role and the mean field theory fails. At this limit, T_c and the gap are governed by strong phase fluctuations. A rough estimate of T_c will be given for $\delta \ll 1$ later. But for large enough $\delta(> 5\%)$ we expect that the mean field theory works. We have chosen a set of parameters (t, U, δ) and calculated the gap as a function of T. The ratio of the zero temperature gap to T_c depends sensitively on the choice of the parameters. In our model the true energy gap is related to the order parameter Δ by

$$\text{Gap} = \frac{3\delta J \Delta/2}{\sqrt{1 + (3\Delta J/2W)^2}} \qquad (13)$$

where $2W$ is the band width ($\sim t\delta$). For some choice of the parameters $2\text{Gap}/T_c$ is close to the BCS value. Some experimental results are needed to get a good idea about t and U. Another interesting feature of our model is the dependence of T_c on δ. The difficulty arises because of lack of information about the metallic screening. In the insulating phase $J = 4t^2/U$ is of order 200 K. In the metallic phase, however, U becomes smaller and t remains essentially unchanged. Thus one expects that J increases. Nevertheless, if we assume a reasonable value for J, we find T_c as a function of δ drops sharply around $\delta \sim 0.15$–0.20. This is easily understood physically: when increasing δ, the band gets wider, $W \sim t\delta$; when W is compatible to the bonding energy J, electron pairs will break up by gaining kinetic energy and superconductivity disappears. One can easily see that T_c becomes very small when $W > J$. Our numerical calculation confirms this (see Fig. 1). We have assumed that the self consistent parameter p is zero in our calculation of T_c. However Wilkins[10] points out that inclusion of p shifts the phase boundary of Fig. 1. The numerical analysis also indicates that T_c is very sensitive to the hopping integral t, large t leads to higher T_c. Therefore within limits, pressure will enhance T_c. Although we have only done at this stage a preliminary study on the superconducting state, our theory seems to be able to explain a wide variety of experimental observations. We leave the thermodynamics and many other details of the model for future investigation. Before concluding the letter, we discuss the small δ limit and Gutzwiller projections.

For small δ the phase fluctuations dominate and T_c is severely suppressed. The motion of singlet pairs in the ground state is not accompanied by any charge flow in the insulator. On the other hand, once we dope the system with holes, the singlet pairs next to the holes can carry charge as they move. In other words, the motion of holes can be thought of as motion of singlet pairs carrying a charge. The effective mass associated with this motion has been shown by Hsu[11] to be of the order of the band electron mass (m^*). Thus we can model the small doping limit as a collection of charged valence bond pair bosons. To get a rough estimate of the T_c associated with the superconductivity of these charged bosons, we use the Bose-Einstein condensation formula: $T_c \sim (2\pi/m^*)(n/2.61)^{2/3}$. So $T_c \sim \delta^{2/3}$. A simple calculation shows that when $\delta < 1\%$, T_c is negligibly small. Then T_c builds up very quickly with δ. When $k_B(T_c)$ exceeds J the modelling in terms of bosons breaks down because valence bond pairs break for $k_B T > J$. Further study is underway to understand the small δ behavior, where screening and localization effects also become important.

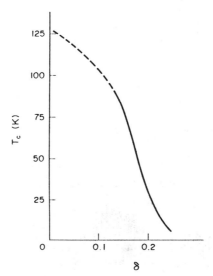

FIGURE 1 T_c vs δ for a choice of parameters $J = 0.05$ eV and band width $2W = 0.4$ eV. The dashed region is the phase fluctuation dominated region, where we expect a drastic reduction in T_c.

Finally about the Gutzwiller projection: We have performed a Gutzwiller projection on the RVB state for the exactly half filled band in one dimension and find that the energy almost coincides with Bethe's ground state energy.[12] The RVB state energy before doing the Gutzwiller projection in one dimension is $-0.608J$ which is lower than the exact energy $\approx -.4431J$. (This is no contradiction to variational principle, because we go outside the Hilbert Space, as for example in the spin wave theory). But after the projection the RVB energy becomes $-.4422J$. So we have demonstrated that in one dimension the energy of our RVB state after the Gutzwiller projection is much closer to the exact result than that of the Néel state. The one dimensional chain has no frustration of course, and the reason that the RVB state has lower energy than Néel state is due to quantum fluctuations. We expect a similar result for the square lattice which has strong quantum fluctuations for not so large U. Additional quantum fluctuation in this two dimensional case could be due to the strong electron phonon interaction and the next nearest neighbor interaction.

The quasiparticle excitation spectrum in the insulator is not qualitatively affected by projecting out the doubly occupied states, since our excitation is chargeless characterizing the spin fluctuations. In the spirit of Fermi liquid theory, the Gutzwiller projection amounts to a mass renormalization which will change the band width of the quasiparticle. The general features of the quasiparticle are preserved qualitatively. We have not been able to perform Gutzwiller projections on our superconducting state satisfactorily for finite δ. Our analysis for the non-half filled band can be taken in the spirit of Fermi liquid theory for superconductivity. The band parameters such as effective mass are Fermi liquid parameters. The on-site Hubbard U induces nearest neighbor singlet correlation between quasiparticles which results in superconductivity. The main aim of this paper has been to show and explain how the antiferromagnetic singlet correlations induce high temperature superconductivity and how quickly this superconductivity is suppressed with high doping.

ACKNOWLEDGEMENTS

We wish to thank E. Abrahams, S. Liang, D. H. Lee, J. Sauls for discussions and J. Yang for help in numerical computations. We also thank Vinay Ambegaokar, Andre Tremble, John Wilkins, Sriram Sastry and a referee for pointing out some corrections to be made in our equations. This work is supported in part by NSF Grant No. DMR-8518163.

RERERENCES

1. J. G. Bednorz and K. A. Muller, *Z. Phys.*, **B64**, 188 (1986); S. Uchida, H. Takagi, K. Kawasaki, S. Tanaka, *Jap. J. A. P.*, letters to appear; R. Cava, B. Batlogg, *et al.*, to appear in *Phys. Rev. Letts.*; C. W. Chu, *et al.*, *Phys. Rev. Letts.*, **58**, 405 (1987); Z. Zhao, *et al.*, *KeXue Tongbao* (China), to be published.
2. P. W. Anderson, *Science*, **235**, 1196 (1987).
3. H. B. Schuttler, *et al.*, preprint; J. D. Jorgensen *et al.*, preprint; L. F. Matheiss preprint; V. Z. Kresin preprint; W. Weber preprint; D. H. Lee, private communication. ; C. M. Varma, S. Schmitt-Rink, and E. Abrahams, private communication and this volume;A. J. Freeman *et al.*, preprint and this volume.
4. P. Ganguly and C. N. R. Rao, *J. Sol. State Chem.*, **53**, 193 (1984). See also K. K. Singh., Ph.D. Thesis, IISc; C. N. R. Rao, K. J. Rao and J. Gopalakrishnan, *Annual Reports C* (Roy. Soc. Chem. London), **233** (1985).
5. P. W. Anderson, *Materials Res. Bull.*, **8**, 153 (1973); P. Fazekas and P. W. Anderson, *Phil. Mag.*, **30**, 432 (1974).
6. M. Noga, preprint.
7. K. Miyake, S. Schmitt-Rink and C. M. Varma, *Phys. Rev. B*, **34**, 6554 (1986).
8. C. Gros, R. Joynt and T. M. Rice, preprint.
9. J. E. Hirsch., *Phys. Rev. Lett.*, **54**, 1317 (1985). Though worked out for extremely small sample, Hirsch seems to have strongly foreshadowed our results. See also this volume.
10. John Wilkins, private communication.
11. T. Hsu, private communication.
12. The Gutzwiller projections of our state in the insulating case can be shown to be identical with that of the simple Fermi gas as worked out by Kaplan, Horsch and Fulde, (*PRL*, **49**, 889 and *J. Phys. C.*, **C16** L1203 (1983)) and in Ref. (7). No corresponding calculations have yet been made in the case with finite T_c.

Andrei E. Ruckenstein, Peter J. Hirschfeld, and J. Appel
Department of Physics
University of California at San Diego
La Jolla, California 92093

Mean-Field Theory of High-T_c Superconductivity: the Superexchange Mechanism[†]

We develop the simplest mean-field theory of an extended Hubbard model in the limit of a large intrasite Coulomb interaction, concentrating on the possibility of superconductivity induced by the superexchange interaction and weakened by the intersite Coulomb repulsion. We calculate the critical temperature and the coherence length as a function of filling, as well as the temperature dependence of the magnetic susceptibility and specific heat. Finally, we comment on the physics of the insulating state at half filling, and mention the probable effects of fluctuations.

Following the pioneering work of Bednorz and Müller[1] and the subsequent discovery of superconductivity above liquid-nitrogen temperatures in the quarternary compound $YBa_2Cu_3O_{4-y}$,[2] a number of theoretical ideas have been proposed to explain this remarkable discovery. Theoretical arguments based on the conventional electron-phonon mechanism appear to rule out critical temperatures much larger than 40 K.[3] Consequently, any other mechanism involving conventional phonons, such as Bose condensation of bipolarons,[4] will most likely lead to a smaller T_c. What remains is the possibility that the superconductivity is due to an unconventional mechanism mediated by Coulomb interactions, perhaps enhanced by the presence of phonons.

[†]Reprinted from *Phys. Rev. B*, **36**, 857 (1987).

One such possibility, discussed by Varma, Schmitt-Rink, and Abrahams,[5] is inxpairing due to charge-transfer excitations (excitons) $Cu^{2+}O^{2-} \rightarrow Cu^+O^-$, which dominate for sufficiently small intrasite Coulomb repulsion. This covalent-metallic picture should be contrasted with the modest covalency and large intrasite Coulomb repulsion characteristic of almost ionic (semi)conductors. The latter underlies Anderson's inxresonating valence bond (RVB) mechanism,[6] in which the superconductivity occurs as a result of spin correlations induced by superexchange between electrons on nearest-neighbor Cu sites.[7]

In this paper we further examine this possibility. We develop a simple mean-field theory of the Hubbard Hamiltonian, which describes spin-$\frac{1}{2}$ fermions on a two-dimensional (2D) square lattice, interacting through intrasite (U) and nearest-neighbor (V) Coulomb repulsions:

$$H = -t \sum_{\substack{(i,m) \\ \sigma}} c_{i\sigma}^\dagger c_{i+m,\sigma} + U \sum_i n_{i\uparrow} n_{i\downarrow} + \frac{V}{2} \sum_{(i,m)} (n_i - 1)(n_{i+m} - 1), \qquad (1)$$

where $n_{i\sigma} = c_{i\sigma}^\dagger c_{i\sigma}$ is the number operator for electrons of spin $\sigma(= \pm 1/2)$ at site i, $n_i = \sum_\sigma n_{i\sigma}$ is the number density, and (i,m) implies unrestricted summation over the nearest neighbors (of site i).

Our discussion is based on the band-structure calculations of Mattheis,[8] which suggest that the relevant physics can be modeled in terms of two-dimensional bands spanned by the O $2p$ and Cu $3d$ orbitals. It is worth noting that the fermion operators $c_{i\sigma}$ in (1) represent electrons localized on the Cu sites. These are the "quasiparticle" operators introduced by Anderson[9] in his theory of superexchange in transition-metal salts, and have the virtue of eliminating explicit reference to ligand electrons. Here we will make use of the quasiparticle picture away from half filling, and we thus rely on the large electron affinity of the oxygen. The effective interaction parameters t, U, and V can be crudely estimated by extrapolating Anderson's discussion of transition-metal salts to our case: We obtain the values $t \sim 0.25$–$0.5\,\text{eV}$, $U \sim 3$–$4\,\text{eV}$, and $V \sim 1$–$2\,\text{eV}$.

Below we assume the large-U limit, and make use of a canonical transformation to eliminate states with doubly occupied sites,[10] yielding

$$H = -t \sum_{\substack{(i,m) \\ \sigma}} b_i f_{i\sigma}^\dagger f_{i+m,\sigma} b_{i+m}^\dagger + \frac{J}{2} \sum_{(i,m)} (\sigma_i \cdot \sigma_{i+m} - n_i n_{i+m})$$

$$-J \sum_{\substack{(i,m,m') \\ \sigma \\ m \neq m'}} b_i (f_{i,\sigma}^\dagger f_{i+m,-\sigma}^\dagger f_{i+m,-\sigma} f_{i+m+m',\sigma}$$

$$- f_{i,\sigma}^\dagger f_{i+m,-\sigma}^\dagger f_{i+m,\sigma} f_{i+m+m',-\sigma}) b_{i+m+m'}^\dagger$$

$$+ \frac{V}{2} \sum_{(i,m)} (n_i - 1)(n_{i+m} - 1) + \sum_i \lambda_i \Big(b_i^\dagger b_i + \sum_\sigma f_{i,\sigma}^\dagger f_{i,\sigma} - 1 \Big) - \mu \sum_{i\sigma} n_{i\sigma}, \qquad (2)$$

where $J = 2t^2/(U - V)$ is the superexchange interaction. Here we have introduced the representation of the fermion operators $c_{i,\sigma}(c_{i,\sigma}^\dagger)$ acting on states with no double occupancy, $c_{i,\sigma} = b_i^\dagger f_{i,\sigma}$, in terms of fermions, $f_{i,\sigma}(f_{i,\sigma}^\dagger)$, and auxiliary Bose fields, $b_i(b_i^\dagger)$. The latter keeps track of the occupancy at a given site, with $b_i^\dagger b_i = 0$ for occupied and $b_i^\dagger b_i = 1$ for unoccupied sites. The condition of no double occupancy (which implies that $\sum_\sigma f_{i,\sigma}^\dagger f_{i,\sigma}$ is at most unity) is incorporated through the constraints $b_i^\dagger b_i + \sum_\sigma f_{i,\sigma}^\dagger f_{i,\sigma} = 1$, enforced independently at each site by the Lagrange multipliers, λ_i. Above, $\sigma_i = \sum_{\sigma\sigma'} f_{i,\sigma}^\dagger \vec{\tau}_{\sigma\sigma'} f_{i,\sigma'}$ and $n_i = \sum_\sigma f_{i,\sigma}^\dagger f_{i,\sigma}$ are the spin and number densities at site i; the sums over m and m' again correspond to unrestricted sums over nearest neighbors. The first and third terms in Eq. (2) represent, respectively, the hopping of electrons, and of pairs of electrons, away from half filling, while the second term gives rise to the superexchange interaction.

We begin by considering the simplest mean-field theory of (2), based on two approximations. First of all, the auxiliary boson is replaced by a filling-dependent c number, while the Lagrange multiplier is taken to be uniform $(\lambda_i = \lambda)$.[11] In this particular case the appropriate procedure leads to a replacement of

$$b_i \to b_i \approx b(1 - n/2)^{-1/2} = [2\delta/(1 + \delta)]^{1/2} \equiv \tilde{\delta}^{1/2}$$

(the last equality follows from the mean field form of the constraint, $b^2 + \sum_{i\sigma}\langle f_{i,\sigma}^\dagger f_{i,\sigma}\rangle/N = b^2 + n = 1$), where $n \equiv \langle n_i \rangle \equiv 1 - \delta \le 1$. This replacement is consistent with the requirement that the mean-field theory applied to a fully spin-polarized partially filled band leads to no mass renormalization.[11] We note that treating the bosons as c numbers amounts to assuming the Bose condensation of the b's. Our second assumption is that, for any value of the filling, the order parameter takes the form

$$\tilde{\Delta}(T, \delta) \equiv \sum_{i,m}\langle b_i^\dagger f_{i,\downarrow} f_{i+m,\uparrow} b_{i+m}^\dagger \rangle/zN$$

$$\approx \tilde{\delta}\sum_{i,m}\langle f_{i,\downarrow} f_{i+m,\uparrow}\rangle/zN \equiv \tilde{\delta}\Delta;$$

$\tilde{\Delta}$ represents the pairing of the physical electrons with opposite spins on nearest-neighbor sites, while Δ can be interpreted as the amplitude of preexisting pairs.

We now proceed with the Hartree-Fock factorization of (2), which leads to the following BCS-like Hamiltonian:

$$H_{\text{BCS}} = \sum_{k,\sigma}\xi(k)f_{k,\sigma}^\dagger f_{k,\sigma}$$

$$- I\sum_k \tau(k)(\Delta^* f_{-k,\downarrow} f_{k,\uparrow} + \text{h.c.}), \qquad (3)$$

where $\tau(k) = 2[\cos(k_x a) + \cos(k_y a)]$ (a is the lattice spacing), $\xi(k) = -\tilde{\delta}t\tau(k) - \mu + \lambda$, and $I \equiv 4J(1 + 3\tilde{\delta}) - V$, Both hopping terms—for single electrons and for pairs—are linear in the concentration of bosons, since both processes require the presence of a single empty site. It is important to note that the fully screened Coulomb repulsion V must appear in the effective interaction I on equal footing with J, since in this case, in contrast with BCS theory, both the pairing mechanism *and* the screening of V occur on the same frequency scale.

The Hamiltonian (3) is diagonalized by the Bogoliubov canonical transformation by introducing the quasiparticle operators, $\gamma_{k0} = u_k f_{k,\uparrow} - v_k f^\dagger_{-k,\downarrow}$ and $\gamma_{k1} = u_k f_{-k,\downarrow} + v_k f^\dagger_{k,\uparrow}$; in thermal equilibrium u_k and v_k can be chosen as real, and are given by

$$u_k^2 = [1 + \xi(k)/E(k)]/2, \quad v_k^2 = [1 - \xi(k)/E(k)]/2,$$

where $E(k) = [\xi^2(k) + I^2\Delta^2\tau^2(k)]^{1/2}$ is the quasiparticle energy; without loss of generality we take $u_k > 0$, and $\mathrm{sgn}(v_k) = \mathrm{sgn}[\tau(k)]$. The self-consistent equations for the gap parameter Δ and the chemical potential $\tilde{\mu}(= \mu - \lambda)$ are then obtained from the defining relations, $\Delta = \sum_k \tau(k)\langle f_{-k,\downarrow} f_{k,\uparrow}\rangle/z$, $n = \sum_{k,\sigma}\langle f^\dagger_{k,\sigma} f_{k,\sigma}\rangle = 1 - \delta$:

$$2z/I = \sum_k \frac{\tau^2(k)}{E(k)} \tanh\left(\frac{\beta E(k)}{2}\right),$$

$$\delta = \sum_k \frac{\xi(i)}{E(k)} \tanh\left(\frac{\beta E(k)}{2}\right). \tag{4}$$

We can now proceed to a brief discussion of our results.

(1) The critical temperature. T_c [obtained by setting $\Delta(T_c) = 0$ in Eq. (4)], starts out with a finite value $T_c(\delta - 0) \approx J - V/2$ (see Fig. 1). Away from half filling, T_c first increases to a maximum determined by the competition between pair hopping, which is linear in $\tilde{\delta}$, and the pair breaking due to single-particle hopping, an effect of order $\tilde{\delta}^2$ for small $\tilde{\delta}$. T_c then decreases rapidly until superconductivity is destroyed through the unbinding of pairs for $\tilde{\delta} \sim I/(2zt)$, where the binding energy I becomes of the order of the effective bandwidth $2zt\delta$. For higher values of $\tilde{\delta}$ the gap equation yields nontrivial solutions for T_c, not shown in Fig. 1. These solutions are, however, ignored as they are associated with values of $[4t + \tilde{\mu}(\delta)] < I$ for which true bound states split off below the bottom of the band. The critical temperature in this regime is always smaller than that for $[4t + \tilde{\mu}(\delta)] \approx I$ and is expected to be a smoothly decreasing function with increasing $\tilde{\delta}$; its value is controlled by the center of mass degrees of freedom of the pairs, and is outside the scope of our BCS-like mean-field theory.

We note that the finite value of T_c at $\delta = 0$ is an artifact of our mean-field approximation. First, unlike the gap parameter Δ the physical order parameter $\tilde{\Delta} \approx \tilde{\delta}\Delta$ vanishes for $\delta = 0$, even within this simple mean-field theory. More importantly, true off-diagonal long-range order implies long-range coherence of the Bose fields b in

$$\tilde{\Delta}(T, \delta) = \sum_{i,m}\langle b^\dagger_i f_{i,\downarrow} f_{i+m,\uparrow} b^\dagger_{i+m}\rangle/zN,$$

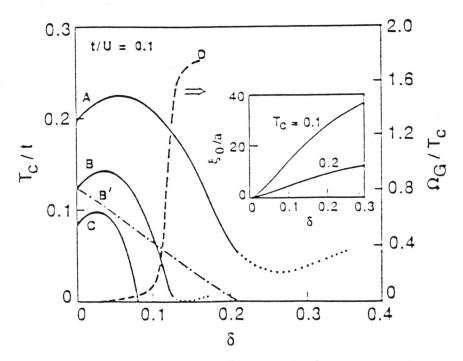

FIGURE 1 Critical temperature (T_c) *vs.* δ for $t/U = 0.1$, with $t/V = \infty$, 3.0, 2.0 for curves A, B, and C, respectively. excitation gap (Ω_G) vs. δ for $t/U = 0.1$ and $t/V = 3.0$ (curve D). inset: $T = 0$ coherence length vs. δ for $T_c = 0.1$ and $T_c = 0.2$. The dotted segments of curves A and B denote the non-BCS-like regime $4t + \tilde{\mu}(\delta) < I$. Curve B' represents T_c vs. δ for d-wave-like pairing with $t/U = 0.1$ and $t/V = 3.0$.

which is clearly absent at $\delta = 0$ due to the "incompressibility" constraint, $b_i^\dagger b_i + \sum_\sigma f_{i,\sigma}^\dagger f_{i,\sigma} = 1$. Close to half filling T_c is determined by the Bose fluctuations. These can be very crudely incorporated into our mean-field theory by explicitly accounting for the thermal depletion of the b condensate, in which case $\tilde{\Delta}(T, \delta) \propto b_0^2(T, \delta)\Delta(T, \delta)$ (b_0^2 is the b condensate fraction). Thus, near $\delta = 0$, T_c is limited by the condensation temperature of the bosons, $T_{BE} \sim 8\tilde{\delta}^v J \Delta^2 a^2$, $v \sim \frac{2}{3}$ in 3D, whereas $v \sim 1$ in 2D; with increasing $\tilde{\delta}$, $T_c(\delta)$ is expected to reach a maximum for $\tilde{\delta} < I/2zt$. This point of view is, however, too naive: the off-diagonal long-range order of the superconducting state is associated with a particular value of a *single* phase variable, namely, that of the physical order parameter $\tilde{\Delta}(T, \delta)$, rather than *two* phases (one of the superconducting gap and one of the b condensate) as implied by the above mean-field picture. These two points of view are reconciled by a precise treatment of the double occupancy constraint, which in turn restores the gauge symmetry with respect to one combination of the phases.

The above discussion applies to the regime in which (a) J is sufficiently large that T_c is nonzero, and (b) the BCS-like pairing theory is valid, *i.e.*, $4t + \tilde{\mu}(\delta) > I$. We note that the results are extremely sensitive to small changes in the parameters.

In fact, for the most realistic estimates quoted above, we find that the second condition is not satisfied, thus requiring a treatment of the crossover between Cooper (*i.e.*, BCS-like) pairing and Bose condensation, a regime outside the scope of this paper. More importantly, for the (t and U) parameter values discussed above we find that superconductivity survives only for unrealistically small values of V, suggesting that even within the mean-field theory help from another attractive interaction (*e.g.*, phonons) may be required to fully stablize the superexchange mechanism. For the sake of illustration, in the figures we have chosen parameter values consistent with both (a) and (b) for the range of fillings specified in the captions.

(2) Coherence length. The absence of coherence in the insulating phase is further substantiated by a calculation of the $T = 0$ coherence length ξ_0 based on the usual derivation of the Ginzburg-Landau free energy. As seen in Fig. 1, ξ_0 vanishes at half filling, as expected in the case of the incompressible ($F_0^s = \infty$) Fermi liquid[12] implied by our mean-field theory. Note that ξ_0 is a measure of the correlation between *different* pairs but that, in the contrast to BCS theory, it does not coincide with the extent of a *single* pair (here taken to be of the order of the lattice spacing).

(3) Excitation spectrum and tunneling density of states. At half filling, our mean-field theory gives a realization of state discussed by Anderson.[6] The Bogoliubov quasiparticle excitations define a "pseudo-Fermi surface" $E(k) = I\Delta|\tau(k)| = 0$ across which $\tau(k)$ changes sign. In close agreement with Anderson's ideas,[6] the low-lying excited states with a fixed average number of particles are of the form $|k', s'; k, s\rangle \equiv \gamma_{k',s'}^\dagger \gamma_{k,s}^\dagger |G\rangle$ ($s, s' = 0, 1$), and correspond to the gapless excitations of a Fermi liquid. The nature of the excitations can be further understood by considering the effect of an arbitrary external potential on $|G\rangle$. More precisely, the coherence factors associated with the coupling of these excitations to external density (or charge) and spin fluctuations are $(u_k, V_k + v_k, u_k)$ and $((u_k, v_k - v_k, u_k)$, respectively. Since at half filling $v_k = \text{sgn}[\tau(k)]u_k = \pm 1/\sqrt{2}$, it is easy to see that whenever k and k' are "on the same or opposite sides of the pseudo-Fermi surface" [*i.e.*, $\text{sgn}\tau(k) - \text{sgn}\tau(k')$ or $\text{sgn}\tau(k) = -\text{sgn}(k')$] the states $|k', s'; k, s\rangle$ represent spinless charge excitations or chargeless spin excitations, respectively. The fact that some excited states carry charge even in the limit of infinite U is easily understood since within our mean-field theory the double occupancy constraint is only satisfied on the average, rather than independently at each site. A proper treatment of the constraints will eliminate all charge fluctuations, but in that case the precise nature of the spin excitations remains to be determined. We have also calculated the single-particle density of states, which is gapless for $\delta = 0$ as expected. Away from $\delta = 0$ an excitation gap, $\Omega_G = I\Delta|\tilde{\mu}|/(I^2\Delta^2 + \tilde{\delta}^2 t^2)^{1/2}$, opens at the true Fermi surface, which no longer coincides with the "pseudo-Fermi surface." An important quantity, accessible in tunneling experiments, is the ratio Ω_G/T_c, which we find to increase with $\tilde{\delta}$ from zero to a value of order of unity for $\tilde{\delta} \sim I/(2zt)$.

(4) Magnetic susceptibility and specific heat. The spin susceptibility and specific heat were calculated from the expressions appropriate for a gas of noninteracting quasiparticles (for illustration we show plots of the specific heat in Fig. 2). In

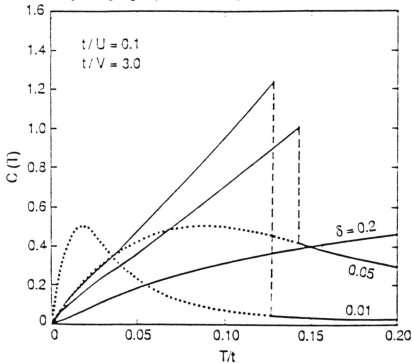

FIGURE 2 Specific heat *vs.* T for $t/U = 0.1$, and $t/V = 3.0$ for $\delta = 0$, 0.05, 0.2. Dotted lines represent normal behavior with $\Delta = 0$.

the "normal state" (*i.e.*, $\Delta = 0$), by increasing δ from zero at a fixed temperature T, we find a cross-over from localized spins fluctuations (at $\delta = 0$), to a Boltzmann gas with effective mass $m_{\text{eff}} \sim 1/\tilde{\delta}$ for $\tilde{\delta} < T/2zt$, and finally to a degenerate non-interacting Fermi gas when $2zt\tilde{\delta} \sim T$. Consistent with the discussion above, for finite Δ and $\delta = 0$, the thermodynamics is essentially that expected of a Fermi liquid with an effective mass $m^*(\delta = 0) \propto (I\Delta a^2)^{-1}$, and crosses over to that of a collection of localized spins for $T > T_c(\delta = 0)$. In the superconducting state ($\Delta \neq 0$, $\delta \neq 0$), the physics is qualitatively similar to that at half filling (for $\Delta \neq 0$), with the associated Fermi-liquid-like behavior for both susceptibility and specific heat. The exponential falloff associated with the gap in the excitation spectrum is visible only at the lowest temperatures, since within the mean-field theory for most fillings $T_c > \Omega_G$. We further note that for reasonable values for the parameters the effective band-width defined by the spin excitations, of order $zI\Delta$, does not imply an appreciable enhancement of either Sommerfeld's γ or the magnetic susceptibility, as one might have expected near a Mott metal-insulator transition.

Above we have discussed the simple mean-field theory of high-T_c superconductivity based on the high U/t limit of an extended Hubbard model. A number of important problems remain open. We have ignored the antiferromagnetic long-range order predicted by the model at half filling. This is sensible only provided the antiferromagnetism is destroyed for sufficiently small δ. The interplay between superconductivity and antiferromagnetism will be considered in a further publication.[13] We have also not addressed the crucial problem of the stability of our mean-field theory with respect to fluctuations, a point particularly worrisome in two dimensions. Order parameter fluctuations, thermal phonons,[14] as well as inelastic electron-electron scattering, should lead to a decrease of T_c. Since the excitation gap is most likely not drastically affected, the ratio Ω_G/T_c will also be increased in the direction of better agreement with experiment.[15] Finally, as already mentioned above, a correct treatment, of the (auxiliary) boson fluctuations is crucial in enforcing the "no double occupancy" constraint, which in turn determines the nature of the physics close to half filling. It is, in fact, not clear that a systematic treatment of the Hubbard model with purely repulsive interactions can lead to a stable superconducting state. We expect, however, that phonons or another exclusively attractive interaction acting together with the superexchange mechanism will lead to a superconducting state, qualitatively similar to that described by the above mean-field theory. In the final stages of writing, we received a paper by Baskaran, Zou, and Anderson[16] which overlaps considerably with the present work, and reaches similar conclusions.

Note added. We have also considered the possibility of d-wave pairing with an order parameter proportional to $d(k) = 2[\cos(k_x a) - \cos(k_y a)]$, the only representation consistent with the nearest-neighbor hopping assumed above. (We note that odd parity pairing does not occur in this model.) The resulting $T_c(\delta)$ is indicated in Fig. 1 as curve B′. We note that for finite δ, $T_c^d < T_c^s$, for a large range of fillings, in contrast to an independent calculation of Kotliar.[17] The latter author ignored the pair hopping term which is important for the stabilization of the s-wave-like state. Kotliar makes, however, the important observation that, since the *s*- and d-wave-like states are degenerate at $\delta = 0$ (at least in the absence of phonons), a mixed state will lead to a gap in the excitation spectrum.

ACKNOWLEDGEMENTS

We acknowledge useful conversations with S. Doniach, D. Haldane, J. Hirsch, M. Inui, R. Klemm, A. Overhauser, S. Schmitt-Rink, L. Sham, E. Siggia, H. Suhl, and P. Wölfle. A.E.R. and P.H. were supported by the Office of Naval Research Grants No. N00014-86-K-0630 and No. N0014-86-WR-24253. P.H. was also supported by ONR Grant No. N00014-82-K-0524.

REFERENCES

1. J. G. Bednorz and K. A. Müller, *Z. Phys. B*, **64**, 188 (1986).

2. M. K. Wu, J. R. Ashburn, C. J. Torng, P. H. Hor, R. L. Meng, L. Gao, Z. J. Haung, Y. Q. Wang, and C. W. Chu, *Phys. Rev. Lett.*, **58**, 908 (1987); J. Z. Sun, D. J. Webb, M. Naito, K. Char, M. R. Hahn, J. W. P. Hsu, A. D. Kent, D. B. Mitzi, B. Oh, M. R. Beasley, T. H. Geballe, R. H. Hammond, and A. Kapitulnik, *ibid.*, **58**, 1574 (1987); C. Politis, J. Geerk, M. Dietrich, B. Obst, and H. L. Luo, *Z. Phys. B*, to be published.
3. W. Weber, *Phys. Rev. Lett.*, **58**, 1371 (1987); **58**, 2154(E) (1987).
4. P. Prelovšek, T. M. Rice, and F. C. Zhang, to be published.
5. C. M. Varma, S. Schmitt-Rink, and E. Abrahams, *Solid State Commun.* **62**, 681 (1987).
6. P. W. Anderson, *Science*, **235**, 1196 (1987).
7. K. Miyake, S. Schmitt-Rink, and C. M. Varma, *Phys. Rev. B*, **34**, 6554 (1986); J. E. Hirsch, *Phys. Rev. Lett.*, **54**, 1317 (1985); M. T. Beal-Monod, C. Borbonnais, and V. Emery *Phys. Rev. B*, **34**, 7716 (1986); D. J. Scalapino, E. Loh, and J. E. Hirsch, *ibid.*, **34**, 8190 (1986); see also, I. A. Privorotskii, *Zh. Eksp. Teor. Fiz.*, **43**, 2255 (1962) [*Sov. Phys. JETP*, **16**, 1593 (1963)].
8. L. F. Mattheis, *Phys. Rev. Lett.*, **58**, 1028 (1987); J. Yu, A J. Freeman, and J. H. Xu, *ibid.*, **58**, 1035 (1987).
9. P. W. Anderson, *Phys. Rev.*, **115**, 2 (1959).
10. J. E. Hirsch, *Phys. Rev. Lett.*, **54**, 1317 (1985); C. Gros, R. Joynt, and T. M. Rice, *Phys. Rev. B*, **36**, 381 (1987). The latter authors, as well as G. Baskaran, Z. Zou and P. W. Anderson, [*Solid State Commun.*, **63**, 973 (1987)], have ignored the pair hopping contribution.
11. G. Kotliar and A. E. Ruckenstein, *Phys. Rev. Lett.*, **57**, 1362 (1986).
12. P. W. Anderson, private communication.
13. M. Inui, S. Doniach, P. Hirschfeld, and A. E. Ruckenstein, unpublished; M. Inui, S. Doniach, P. Hirschfeld, and A. E. Ruckenstein, to be published.
14. J. Appel, *Phys. Rev. Lett.*, **21**, 1164 (1968).
15. M. D. Kirk, D. P. E. Smith, D. B. Mitzi, J. Z. Sun, D. J. Webb, K. Char, M. R. Hahn, M. Naito, B. Oh, M. R. Beasley, T. H. Geballe, R. H. Hammond, A. Kapitulnik, and C. F. Quate, *Phys. Rev. B*, **35**, 8850 (1987).
16. G. Baskaran, Z. Zou, and P. W. Anderson, *Solid State Commun.*, **63**, 973 (1987) and this volume.
17. G. Kotliar, to be published.

Ian Affleck
Physics Department
University of British Columbia
Vancouver, BC V6T 2A6

and

J. Brad Marston
Joseph Henry Laboratories of Physics
Jadwin Hall, Princeton University
Princeton, New Jersey 08544

The Large-*N* Limit of the Hubbard Model: a New Mean-Field Theory for High-T_c Superconductors[†]

In order to gain insight into the behavior of the Hubbard model and its relevance to the high-T_c superconductors, an SU(n)-invariant generalization of the Hubbard model is defined and solved in the large-*n* limit. Orthorhombic and gapless tetragonal phases are found separated by an electronically driven transition. Enhancement of the tendency towards superconductivity is seen near the structural transition.

Anderson[1] has proposed that the correct model to describe the CuO_2 planes in the high-T_c superconductors is the two-dimensional Hubbard model, *without phonons*: only the electrons in the outer *d*-orbitals of the Cu^{++} ions are retained, and the O^{--} ions are ignored entirely. The electron density is believed to be 1 for pure La_2CuO_4, and to decrease with Sr or Ba doping. Adjusting the O-deficiency in $YBa_2Cu_3O_{7-x}$ should adjust the filling factor in this case as well, although in a less obvious way (due to the presence of the CuO chains). It is well-known that in the limit $U/|t| \gg 1$, with a half-filled band the model reduces to the Heisenberg antiferromagnet, with no charge transport. Estimates suggest that $U \approx 10\,eV$, $t \approx 1\,eV$, and indeed undoped La_2CuO_4 seems to be an insulator and to be well described

[†]Reprinted from *Phys. Rev.B* (in press)(1988

by the Heisenberg model (although the peculiar dependence on O-deficiency is not very well understood). The doped compound $La_{2-\delta}Sr_\delta CuO_4$, with electron density $1-\delta$, is superconducting for $\delta \approx .15$.[2] Meanwhile an orthorhombic tetragonal structural transition occurs at approximately the same value of δ, with the orthorhombic phase being stable at lower δ.[2]

It is clearly very important to understand whether the Hubbard model can produce this behavior. Anderson and collaborators[3,4] have developed a language and mean field theory for understanding these phenomena based on the notion of "resonating valence bonds." Almost nothing is known rigorously about either the $s = 1/2$ Heisenberg model or the Hubbard model on a square lattice.

Diagonalization of small systems[5] suggests that the Heisenberg model Néel orders at $T = 0$. Indeed Néel order has been observed in some La_2CuO_4 samples.[6] However recent experimental studies[7] of the two-dimensional magnetic correlations suggest that the order is purely a three-dimensional effect and that the two-dimensional system might *not* order, even at $T = 0$.

Another relevant issue is the nature of the gap for spin excitations. Experimental indications of a linear low-temperature specific heat have been reported,[8] suggesting a gapless Fermi surface as was predicted by RVB theory.[3,4] A theorem due to Lieb, Schultz and Mattis[9] seems to imply[10] that either gapless excitations or broken translational symmetry occurs.

In this letter we attempt to shed light on these issues by applying a systematic approximation to the Heisenberg and Hubbard models based on the large-n limit.[11] We replace the two spin components of the electron by n "flavor" components, and let n become large. The Heisenberg interaction simply exchanges the flavors of electrons on neighboring sites: $\mathbf{S}_i \cdot \mathbf{S}_j = (1/2)c_i^{+\alpha}c_{i\beta}c_j^{+\beta}c_{j\alpha}+$ constant. (Repeated flavor indices, α, β are summed from 1 to n). To obtain the analogue of a spin system we must freeze the charge excitations by projecting out states with a definite electron number on each site. There are two natural ways of extending the projection made for $n = 2$: we may fix 1 electron on the even sub-lattice and $n - 1$ on the odd sub-lattice. Alternatively (for n even) we may fix $n/2$ on all sites.

In the former case the ground state can be constructed out of nearest neighbor valence bonds. A valence bond, for general n, is the state in which the flavors of a particle and hole are summed over: $|_\alpha^\alpha\rangle \equiv c_i^{+\alpha}c_{j\alpha}|0\rangle$, where $|0\rangle$ has no electrons on site i and all n on site j. If we divide H by n so that the energy of a valence bond is -1, then the probability for H to break valence bonds becomes $O(1/n)$: $H_{23}|_{\alpha\beta}^{\alpha\beta}\rangle = (-1/n)|_{\beta\alpha}^{\alpha\beta}\rangle$. Thus large n suppresses the fluctuation of the valence bonds [while destabilizing the SU(n) breaking Néel phase]. Consequently *any* arrangement of *nearest neighbor* valence bonds is a ground state in the large-n limit. In one dimension this simply leads to two dimerized ground states.[11] In two dimensions at next order in $1/n$, the various valence-bond configurations mix, and some linear combination should be the true ground-state.

If we choose $n/2$ electrons on all sites, then singlet states have $n/2$ valence bonds emanating from each point. Thus there will be, in general, a large number of valence bonds on each link. The probability for H to break a valence bond is not small in this case, but moving a single valence bond has a small effect on the state. Fluctuations in the number of valence bonds on each link become suppressed at large n. This system has the full translational symmetry of the original problem, whereas in the previous version the two sublattices are inequivalent. Consequently the Lieb-Schultz-Mattis theorem implying the existence of a low-energy excitation works in the latter but not the former case.[11] We will focus on the more symmetric case here.

We solve the model at large-n by a path-integral method. This allows us to introduce the valence bond creation operator, $\chi_{ij} \equiv c_i^{+\alpha} c_{j\alpha}$, by a Hubbard-Stratonovich transformation. At large n we may ignore its fluctuations. The (imaginary time) Lagrangian is written

$$\mathcal{L} = \Sigma_j \left[c_j^{+\alpha}(d/d\tau)c_{j\alpha} + i\phi_j(c_j^{\alpha+}c_{j\alpha} - n/2) \right]$$
$$+ \Sigma_{(ij)} \left[(n/4J)|\chi_{ij}|^2 + (c_j^{+\alpha}c_{i\alpha}\chi_{ij} + \text{h.c.}) \right].$$

Here the Lagrange multiplier field, ϕ, is introduced to enforce the constraint on the number of electrons on each site. The electron operators appear quadratically in $\mathcal{L}$ and are integrated over freely. Thus we may integrate out the n flavors of electrons and obtain an effective action for χ_{ij} and ϕ_j which is proportional to n. Under these circumstances the physics is normally controlled by the lowest action saddle-point of this effective action. Since the saddle-point describes the thermal equilibrium state, we expect it to be time-independent. A subtlety arises because of the gauge invariance of $\mathcal{L}$. The exact conservation of the particle number at *each site* implies invariance under phase rotations of the c_j's with arbitrary space and time dependence,[12] $c_j(\tau) \rightarrow \exp[i\theta_j(\tau)]c_j(\tau)$. ϕ and χ transform as $\phi_j \rightarrow \phi_j - d\theta_j/d\tau$. $\chi_{jk} \rightarrow \chi_{jk}\exp[i(\theta_j - \theta_k)]$. Thus the phases of $\chi_{jk} = |\chi_{jk}|\exp(i\theta_{jk})$ act as spatial components of the gauge field, and ϕ as the time component. The sum of the θ_{jk}'s around an elementary plaquette is gauge invariant; it corresponds to the magnetic flux through the plaquette. Compact gauge symmetries are not broken in lattice gauge theories. Gauge-equivalent saddle-points must be averaged over, so that only gauge-invariant quantities have non-zero expectation values. In particular, $\chi_{jk}^+\chi_{jk} \propto \mathbf{S}_j \cdot \mathbf{S}_k +$ constant, may be non-zero.

Assuming a time-independent saddle-point, the electrons simply propagate in the classical ϕ_j, χ_{jk} fields. χ_{jk} acts as a hopping probability and $i\phi_j$ is an on-site potential. Denoting the electron energy levels in these fields $E_a(\phi, \chi)$, the ground-state energy of the system is $\varepsilon(\phi, \chi) = n \left\{ (1/4J)\Sigma_{(jk)}|\chi_{jk}|^2 - (i/2)\Sigma_j\phi_j - \Sigma_a E_a \right\}$ where the last sum is performed up to the Fermi surface for half-filling. We would like to find the global minimum of $\varepsilon(\phi, \chi)$. Due to the non-locality of ε this is not a trivial problem. However, we expect on physical grounds that the minimum will have a reasonably high degree of symmetry. We will restrict our search for the global minimum to configurations corresponding to a $\sqrt{2}$ unit cell as observed in the orthorhombic and Néel phase. Thus ϕ may have different values on the even and odd sublattices, and there are four different values of χ (Fig. 1). ε depends on the phases of the χ's only via their sum, the flux. Fourier transforming the even and odd sub-lattice electrons separately, we obtain a dispersion relation with two branches:

$$E_\pm(k_x, k_y) = (\phi_e + \phi_0)/2 \pm \left[|\chi_1 \exp(ik_x) + \chi_2^* \exp(ik_y)\right.$$
$$\left. + \chi_3 \exp(-ik_x) + \chi_4^* \exp(-ik_y)|^2 - (\phi_e - \phi_0)^2/4\right]^{(1/2)} \quad (1)$$

The term linear in $(\phi_e + \phi_0)$ cancels out in ε (by gauge invariance) and $(\phi_e - \phi_0)$ is zero at the minimum.

We have done a computer search for minima of $\varepsilon(\chi_i)$ finding two locally stable minima:

1. Peierls phase—only one of the four χ's is non-zero. Each spin forms a dimer with one of its nearest neighbors; hence the electronic spectrum is completely localized, with a gap.
2. Flux phase—all four χ's have equal *magnitude* and the sum of the four phases, the flux is π. There is only one gauge-inequivalent state of this kind. Choosing the gauge in which all the χ_i's are equal, the electronic spectrum becomes $E(\mathbf{k}) = \pm|2\chi| \left[\cos^2 k_x + \cos^2 k_y\right]^{1/2}$. Note that the gap only vanishes at $\mathbf{k} = (\pm\pi/2, \pm\pi/2)$. The constraint on the number of particles on each site implies that only particle-hole excitations are permitted with gapless modes at $(0,0)$, $(\pi,0)$, $(0,\pi)$, (π,π).

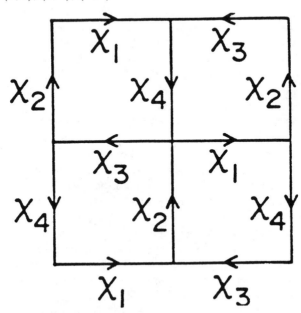

FIGURE 1 Definition of the four χ's consistent with invariance under translation across a diagonal.

These results are consistent with the Lieb-Schultz-Mattis theorem.[9,10] In the Peierls phase the symmetry of translation by one site is spontaneously broken, and in the flux phase there are gapless excitations of wave-number $(\pi, 0)$.

We have, in fact checked that these solutions are local minima with respect to arbitrary *space-dependent* variations of ϕ and χ. We find that the Peierls phase is slightly lower in energy.

We now generalize to the Heisenberg-Hubbard model with finite U/t, arbitrary filling and an explicit Heisenberg interaction:

$$\mathcal{L} = \Sigma_j \left[(n/2U)\phi_j^2 + c_j^{+\alpha}(d/d\tau)c_{j\alpha} + i\phi_j(c_j^{\alpha+}c_{j\alpha} - n/2) \right]$$
$$+ \Sigma_{(ij)} \left\{ (n/4J)|\chi_{ij}|^2 + \left[c_j^{+\alpha}c_{i\alpha}(\chi_{ij} + t) + \text{h.c.} \right] \right\}.$$

Taking J to 0 gives the ordinary Hubbard model, and taking U to infinity (for fixed t and half-filling) gives the Heisenberg model. Again, in the large-n limit, the fluctuations of ϕ and χ are suppressed, so these fields become classical. The gauge-invariance of the Heisenberg model is broken, leaving only the global $U(1)$ charge symmetry [together with SU(n)], under which both ϕ and χ are invariant. Thus, in the large-n limit, neither the $U(1)$ or SU(n) symmetries can be broken; neither antiferromagnetism nor superconductivity occurs. The impossibility of singlet superconductivity at large-n is associated with the fact that n electrons are required to form a singlet, and such a Cooper n-tuplet is hard to form when n is large. Only the spatial symmetries can be spontaneously broken in the large-n limit, leading to some type of charge-density-wave (CDW) state. Of course, as n is lowered, breaking of the continuous symmetries might occur.

The phase of χ is related to the superconducting order parameter. For the $n = 2$ case, we may write $\chi_{jk}^2 = (c_i^{+\alpha}c_{k\alpha})^2 = -2(c_j^{+\uparrow}c_j^{+\downarrow})(c_{k\uparrow}c_{k\downarrow}) = -2\Delta_j^+\Delta_k$, where Δ_i is the superconductivity order-parameter. Introducing its phase, $\Delta_j = |\Delta_j|\exp(i\theta_j)$, we see that the phase of $\chi_{jk} = |\chi_{jk}|\exp(i\theta_{jk})$, is one-half the *difference* of the superconductivity order-parameters on sites j and k; $\theta_{jk} = (\theta_k - \theta_j)/2$. For general n, Δ_j is a product of n electron annihilation operators and the phases of Δ_j and χ_{jk} are related by $\theta_{jk} = (\theta_k - \theta_j)/n$. In the large-$n$ limit (for any finite t) θ_{jk} has some classical value $\overline{\theta}_{jk}$ and fluctuations are suppressed; explicitly the effective action contains a term $\kappa n(\theta_{jk} - \overline{\theta}_{jk})^2$, where κ is a constant, depending on t, δ etc. However the suppression of phase fluctuations of θ_j is small, $\kappa(\theta_j - \theta_k - \overline{\theta}_j + \overline{\theta}_k)^2/n$. Thus χ can have a non-zero expectation value without Δ having one. However, as n is decreased towards 2, the suppression of fluctuations of θ_j increases and eventually long-range-order may set in. The dependence of the stiffness, κ, on t and δ gives an indication of the range of parameters where superconductivity is most enhanced, due to the suppression of phase fluctuations of the order parameter.

There are two types of CDW order that can occur depending on whether the magnitude of χ or ϕ has an alternating expectation value. Since $i\phi_j = U(n_j - n/2)$, alternation in ϕ indicates a site-centered CDW. On the other hand, χ measures the number of valence bonds on a given link. Strictly speaking, in our model, the electrons are forced to live on the sites. However, taking a more realistic point of view, we may consider a pair of electrons forming a valence bond on some link to reside in the vicinity of the O^{--} ion situated at the midpoint of that link. Thus alternation in χ indicates a bond-centered CDW.

Finally, a non-zero expectation value for θ_{jk} indicates a diamagnetic current along the link: $\langle c_j^+ c_k - c_k^+ c_j \rangle = \chi_{jk} - \chi_{jk}^*$.

The electrons now propagate with a hopping term $(t + \chi_{jk})$ between sites j and k, and an on-site potential $i\phi_j$. Making the same symmetry assumption as above, the electron energies are again given by Eq. (1) with χ_{jk} replaced by $\chi_{jk} + t$. Again, for positive U, ϕ is zero. On the other hand, for negative U the term $(n/2U)\Sigma_j \phi_j^2$ favors a non-zero ϕ and a site-centered CDW might develop. At $J = 0$, we expect $\chi = 0$ so that the solution of the non-interacting case emerges. However a Fermi surface instability resulting from the perfect nesting for a square lattice nearest neighbor system, implies a logarithmic divergence of the wave-vector integral contributing to $\partial \varepsilon / \partial |\chi|^2|_0$. Thus (exponentially small) dimerization sets in at arbitrarily small J.

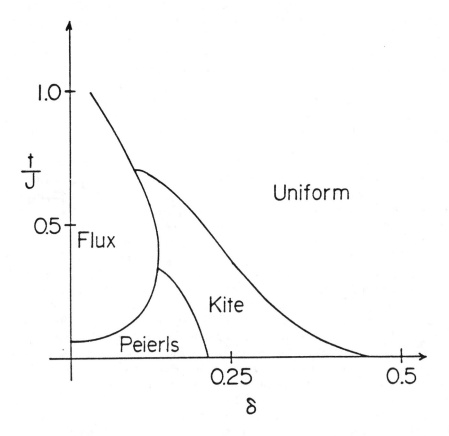

FIGURE 2 Approximate phase diagram of the $n \to \infty$ Heisenberg-Hubbard model as a function of Heisenberg exchange constant and doping.

A computer search for minima reveals the phase diagram of Fig. 2. For small δ we find the same two phases discussed above (but in general in the Peierls phase the smaller three χ's are equal but non-zero and in the flux phase the flux is not necessarily π). Two other phases also appear at larger δ:

3. Uniform phase: all four χ's are equal and positive. Thus χ simply acts as an addition to t, and the spectrum is that of free fermions. No symmetries are broken.

4. Kite phase: $\chi_1 = \chi_2 \neq \chi_3 = \chi_4$ (or $\chi_1 = \chi_4 \neq \chi_3 = \chi_2$) with $|\chi_1| \neq |\chi_3|$. There are equal numbers of valence bonds on alternating zig-zagging diagonal lines through the lattice. The Fermi surface is gapless in all phases for $\delta \neq 0$.

The linearly vanishing gap at discrete points in momentum space, seen in the flux phase at $\delta = 0$, corresponds to $(2 + 1)$-dimensional relativistic behavior, the low-energy sector should be a massless quantum field theory. As the temperature is lowered sharp peaks should occur in the neutron scattering cross-section for wave-number near $(0, \pi)$ and $(\pi, 0)$ as observed.[7] This phase has diamagnetic currents flowing around the elementary plaquettes in an antiferromagnetic fashion (alternating sign).

The kite phase has an interesting correspondence with the observed orthorhombic distortion in La_2CuO_4. The main distortion is a buckling of the O atoms out of the planes, with alternating diagonal line of O's distorting in the same direction. This is precisely the response that the O's would make to the kite phase bond-centered CDW. There should be a net negative charge excess on the O atoms with higher valence bond probabilities (χ_1 and χ_2) and an excess positive charge on the other O atoms (χ_3 and χ_4). Coulomb repulsion between the O's should then cause them to move off the plane in alternating direction as we move along a zig-zag line of O's with the same charge excess. (A somewhat related explanation of the orthorhombic distortion was proposed by Anderson, et al.[4]) However, there may be a problem with this interpretation. A distortion of the Cu lattice should also occur in response to the CDW. The observed distortion corresponds to the tilting of the square plaquette into a rhombus (all sides remain equal in length). However the kite phase CDW should result in the Cu's along alternating zig-zag's drawing closer together or farther apart. This results in a distortion of the square into a "kite," i.e., a four-sided figure with pairs of adjacent sides equal in length.

The stiffness $\kappa \to 0$ as $\delta \to 1$, since the phase dependence of ε comes entirely from the sum, over occupied electron states, which goes to zero in this limit. We find a dramatic drop in κ at the kite-uniform phase transition, corresponding to the observation in $La_{2-\delta}Sr_\delta CuO_4$ that T_c rapidly goes to zero as the doping is increased to drive the system into the tetragonal phase.

More details will be given elsewhere.

ACKNOWLEDGEMENTS

We thank P. W. Anderson, G. Baskaran, and E. Fradkin for useful conversations.

REFERENCES

1. P. W. Anderson, *Science* **235**, 1196 (1987).
2. J. D. Jorgensen, H.-B. Schuttler, D. G. Hinks, D. W. Capone II, K. Zhang, M. B. Brodsky, and D. J. Scalapino, *Phys. Rev. Lett.*, **58**, 1024 (1987).
3. P. W. Anderson, *Science*, **235**, 1196 (1987); G. Baskaran, Z. Zou, and P. W. Anderson, *Solid State Communications*, **63**, 973 (1987) and this volume; P. W. Anderson, G. Baskaran, and Z. Zou, to be published; Z. Zou and P. W. Anderson, *Phys. Rev.B* (in press) and this volume. G. Baskaran and P. W. Anderson, Princeton preprint.
4. P. W. Anderson, G. Baskaran, Z. Zou, and T. Hsu, *Phys. Rev. Lett.*, **58**, 2790 (1987).
5. J. Oitmaa and D. D. Betts, *Can. J. Phys.*, **56**, 897 (1978).
6. D. Vaknin, S. K. Sinha, D. E. Moncton, D. C. Johnston, J. M. Newsam, C. R. Safinya, and H. E. King, Jr., *Phys. Rev. Lett.*, **58**, 2802 (1987).
7. G. Shirane, Y. Endoh, R. J. Birgeneau, M. A. Kastner, Y. Hidaka, M. Oda, M. Suzuki, and T. Murakami (1987) preprint.
8. N. E. Phillips, R. A. Fisher, S. E. Lacy, C. Marcenat, J. A. Olsen, W. K. Ham, and A. M. Stacy (1987), L. B. L. preprint; E. Airniebl, J. Willis, J. Thompson, C. Huang and J. Smith, (1987) Los Alamos preprint.
9. E. Lieb, T. Schultz and D. Mattis, *Ann. Phys. (N.Y.)*, **16**, 407 (1961).
10. I. Affleck, unpublished.
11. I. Affleck, *Phys. Rev. Lett.*, **54**, 966 (1987).
12. I. Affleck and E. Lieb, *Lett. Math. Phys.*, **12**, 57 (1986).
13. The invariance under *time*-dependent gauge transformations was pointed out to us by E. Fradkin.

Steven A. Kivelson, Daniel S. Rokhsar, and James P. Sethna
Institute for Theoretical Physics
University of California
Santa Barbara, California 93106

Topology of the Resonating Valence-Bond State: Solitons and High-T_c Superconductivity[†]

We study the topological order in the resonating valence-bond state. The elementary excitations have reversed charge-statistics relations: There are neutral spin-$\frac{1}{2}$ fermions and charge $\pm e$ spinless bosons, analogous to the solitons in polyacetylene. The charged excitations are very light, and form a degenerate Bose gas even at high temperatures. We discuss this model in the context of the recently discovered oxide superconductors.

In a recent paper, Anderson[1] proposed that La_2CuO_4 is in a "resonating valence-bond" (RVB) state, and that the high-T_c superconductivity observed in the material[2] when doped with Ba or Sr must be understood in terms of this unusual insulating state. We show here that the resonating valence-bond state has topological long-range order, and characterize its topological excitations. We propose an exotic mechanism for superconductivity.

Our paper draws three main conclusions. (1) We argue that electron-phonon interactions can stabilize a resonating valence-bond state. A highly simplified schematic of this state is shown in Fig. 1(a). The state does not possess a broken symmetry, but on a bipartite lattice it has a topological long-range order. It has a gap

[†]Reprinted from *Phys. Rev. B*, **35**, 8865 (1987).

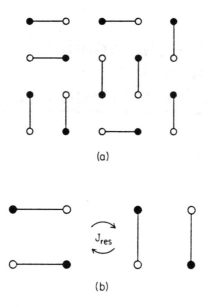

(a)

(b)

FIGURE 1 (a) To first approximation, the resonating valence-bond state is a coherent superposition of states like the one depicted here. Lines denote strong "valence" bonds, across which the electrons form singlet pairs; each site is strongly bonded to precisely one neighbor. The real RVB state is "dressed" by virtual soliton-antisoliton pairs. (b) An elementary resonance between two valence-bond configurations. This resonance lowers the energy of the superposition by J_{res}.

to both spin and charge excitations. (2) The elementary electronic excitations[3] in the RVB state have reversed charge statistics and charge-spin relations: There are neutral spin-$\frac{1}{2}$ fermions and charge $\pm e$ spinless bosons. At finite temperatures, these topological solitons destroy the long-range order (see Fig. 2). (3) The bosons have a mass on the order of the electron mass, and a binding energy set by electronic energy scales, so they will exist and be highly degenerate even at high temperatures. Naively, one might think that the condensation of charge $\pm e$ bosons would lead to flux quantization in units of hc/e. However, the solitons are topological, and hence can only be created or destroyed in pairs; this seems to imply that if flux quantization occurs, the quantum is $hc/2e$, consistent with Yang's[4] general analysis of off-diagonal long-range order. It is plausible that this RVB state occurs in Sr- or Ba-doped La_2CuO_4, and that in the doped material condensation of the charge-e bosons could occur at high temperatures.

The active electronic states in La_2CuO_4 are associated with the two-dimensional square lattice of Cu ions. Each Cu has one active orbital and (in the undoped insulator) has one associated electron. Exchange between Cu ions is mediated by the oxygen atom midway between them; there is thought to be a large Coulomb energy associated with placing two electrons on the same copper. The natural model for this material is the half-filled Hubbard model. There is a structural transition in the insulating state at $533\,\text{K}$,[5] in which the intermediate oxygen atoms buckle out of

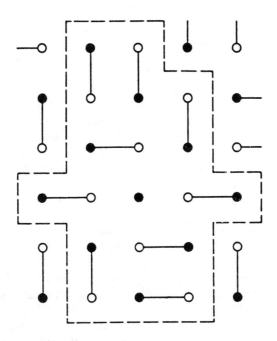

FIGURE 2 The existence of a topological defect (here, a black soliton) can be deduced from a large loop enclosing it.

the plane; initial experiments suggest[6] that the superconducting state is associated with the suppression of this lattice distortion. Thus we are led to look at Hubbard-Peierls models for phases without broken translational symmetry but with a gap to charged excitations. (Our analysis of the statistics of the elementary excitations is, however, independent of the mechanism by which the RVB state is stabilized.)

Fig. 1(a) shows a highly simplified schematic of the resonating valence-bond state, which should be interpreted as a snapshot in a path integral. Each lattice site is occupied by a single electron (large Hubbard U), and participates in one singlet "valence bond" with a nearest-neighbor site. Associated with each valence bond is a lattice deformation which increases the hopping energy across that bond; these phonons stabilize the RVB state with respect to the Néel state (which cannot take advantage of lattice deformations to lower its energy). The RVB state is a coherent superposition of the various arrangements of valence bonds; the resonance energy from tunneling between these arrangements stabilizes the RVB phase with respect to the spin-Peierls phase (which is a crystalline arrangement of bonds).

To examine the competition between the Néel and the RVB states, consider the Heisenberg Hamiltonian (describing the Hubbard model in the large U limit) including the effects of lattice deformations:

$$H = \sum_{\langle i,j \rangle} \left(\frac{4(t_0 - \alpha u_{ij})^2}{U} \mathbf{S}_i \cdot \mathbf{S}_j + \frac{K}{2}(\mathbf{u}_i - \mathbf{u}_j)^2 \right), \qquad (1)$$

where $\mathbf{u}_i$ is the displacement of the ith atom, u_{ij} is the change in the length of the bond between sites i and j, t_0 is the electron hopping matrix element, α is the electron-phonon coupling constant, U is the on-site repulsion between electrons, and K is the spring constant. The naive energy for the Néel antiferromagnetic state (*i.e.*, ignoring quantum corrections) is $-2t_0^2/U$ per site. The naive energy for our RVB state is obtained by balancing the strain energy against the gain in electronic energy; the RVB energy is lower than the Néel state when $\alpha^2/KU > \frac{1}{9}$, at which point the strong-bond hopping matrix element is $\frac{4}{3}t_0$. Quantum corrections (spin waves versus resonance) further favor the RVB state.[7,8]

The RVB state is a quantum liquid of valence bonds.[7] Under what conditions is it lower in energy than crystalline arrangements of valence bonds (*i.e.*, spin-Peierls states)? The most natural spin-Peierls state looks like a stack of antiferromagnetically aligned polyacetylene chains: All bonds are aligned along the same axis, with no next-nearest-neighbor bond pairs. Presumably, this crystalline arrangement minimizes the energy for large ion mass, but without breaking bonds it cannot resonate with any other valence-bond configuration. A typical valence-bond configuration has a large density of next-nearest-neighbor bond pairs. These pairs can resonate between horizontal and vertical configurations (Fig. 2), which lowers the energy of the appropriate superposition by an effective tunnel splitting J_{res}. If the difference in energy density between the spin-Peierls state and a single typical valence-bond configuration is less than order J_{res}, the RVB phase will be preferred. In the adiabatic limit, a simple WKB (Wentzel-Kramers-Brillouin) estimate (ions of mass M tunneling a distance u through a barrier of height $\alpha ut/U \gtrsim \hbar\omega^*$) gives

$$J_{\mathrm{res}} \approx \omega^* \exp\left[-A(t^2/U)/(\hbar\omega^*)\right],$$

where A is a constant of order unity and $\omega^* = \sqrt{\alpha^2/KU}\sqrt{K/M}$ is the renormalized phonon frequency.[9] (This estimate suggests J_{res} is small. In the absence of phonons, J_{res} would be of order t^2/U. Doping can also stabilize[10] the RVB state by a delocalization resonance energy of order xt_0, where x is the soliton density.)

We can see already a schematic of the RVB state's topological long-range order. Color the lattice sites alternately black and red, in a checkerboard pattern; each bond connects two sites—one black, one red. Consider a large loop which cuts no bonds. For a perfect valence-bond configuration,[11] the number of enclosed red and black squares must be the same, independent of the size of the loop. Unbonded sites (dangling bonds) are the topological defects. So long as the loop passes through perfect regions, the difference between the number of black and red squares enclosed is equal to the difference between the number of black and red defects (Fig. 2):

Dangling bonds (free spins) on black squares act as antiparticles to dangling bonds on red squares.

These dangling bonds are entirely analogous to the neutral solitons in poly-acetylene.[9] They have spin $\frac{1}{2}$ and charge 0, are created in pairs by breaking a bond, and (because the RVB state has no crystalline long-range order) act as free particles. As in polyacetylene, the defect will presumably extend over several sites, and be quite mobile. Upon doping, the added electrons or holes will bind to the free spins, forming a charged soliton. (In the absence of electron-electron interactions, charge conjugation symmetry implies that a dangling bond has a midgap state; the charged soliton binding energy is then of order $2\alpha u$. With interactions, these states will split apart, but the extra electron will presumably prefer to avoid breaking an extra bond, giving a binding energy of order $2t^2/U$.) The charged defect has spin 0 and charge $\pm e$. The charged defect will certainly delocalize over several sites; its size R is determined by balancing the delocalization energy (of order $-t[1 - (a/R)^2]$) against the exchange energy [of order $(t^2/U)(R/a)^d$ per site], so $R/a \sim (U/t)^{1/(d+2)}$.

If the many-body wave function has a quasiparticle interpretation, the statistics of the quasiparticles can be determined by considering the transformation of the wave function under the exchange of two solitons as follows. Turn on an external potential which localizes the solitons near point Q_0 and R_0. Schematically, the quasiparticle wave function should be approximately of the form

$$\Psi(Q, R) = [\Phi(Q - Q_0)\Phi(R - R_0) \pm \Phi(Q - R_0)\Phi(R - Q_0)]/\sqrt{2},$$

where Q and R are the quasiparticle coordinates. By slowly varying this external potential the quasiparticles can be moved adiabatically along a path which ex-changes Q_0 and R_0 [Fig. 3(c)]; before and after exchange the Hamiltonian is the same. From the net sign change of the wave function we have determined that the charged solitons are bosons and the neutral solitons are fermions, as described in Fig. 3.

How are these consideration related to high-T_c superconductivity? First, the scale of the binding energy of the elementary bosons is no longer set by the phonon Debye frequency, but rather by an electronic energy, $\sim 2t_0^2/U$. Second, just as for solitons in polyacetylene,[9] the effective mass of the bosons can be very small, making for large quantum effects.[13] In two dimensions,[14] the adiabatic effective mass of a soliton is proportional to the square of the lattice deformation: $M^*/M \approx (u/a)^2$. To estimate M^*, we use parameters for La$_2$CuO$_4$. We take $t_0 = 0.5$ eV and $a \sim$ 3 eV/Å from band-structure calculations.[15] These parameters imply that a lattice displacement $u_c = t_0/(6\alpha) = 0.03$ Å is needed to stabilize the RVB state with respect to the Néel state. The Cu-Cu distance is $a = 3.79$ Å, so at critical coupling $M^*/M \approx 5 \times 10^{-5}$. Thus, as in polyacetylene, the soliton mass is comparable to an electron mass; at optimal doping densities bosons of this mass would still be highly degenerate at the measured T_c. The large binding energy and degeneracy temperature of our charged solitons, both several hundred K, make this a promising starting point for a complete theory.[16]

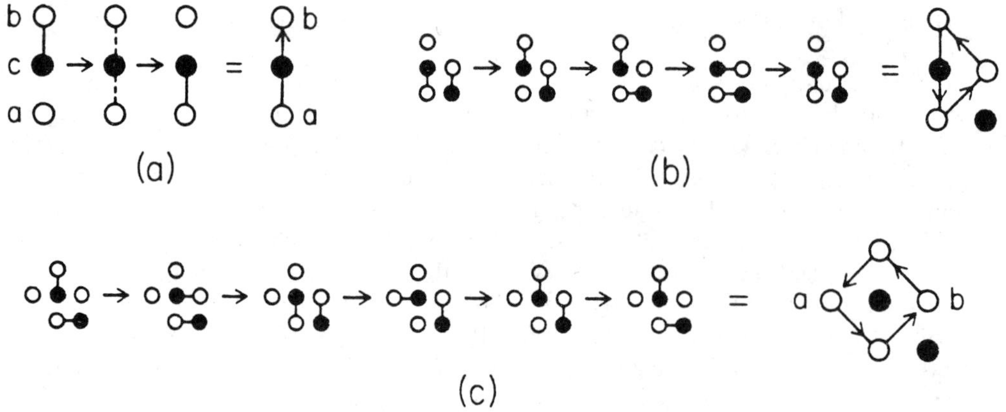

(a)

(b)

(c)

FIGURE 3 Quasiparticle statistics: Statistics are usually determined by the change in the sign of the wave function when the identities of two particles are permuted. Statistics can also be determined by adiabatically dragging two identical quasiparticles along a path which exchanges their positions (Ref. 12), and examining the resulting change in the phase of the many-body wave function. (a) An elementary step which adiabatically transports a soliton from site a to site b. The electron hopping matrix elements t_{ac} and t_{cb} are slowly changed from their initial values $(t_{cb} \gg t_{ac})$ to their final values $(t_{cb} \ll t_{ac})$; this moves the valence bond from cb to ac and the soliton from site a to site b. The sign change in the wave function is unambiguous only if the solitons transverse a closed loop. However, keeping the wave function real and using the most natural sign convention, we find that charged solitons pick up a $+1$ per move and neutral solitons pick up a -1 per move (calculated analytically for $U = 0$ and numerically for the $U \neq 0$ Hubbard model). (b) Transporting *single* neutral solitons around loops produces phase changes. We carry a single red soliton around a closed path which encloses one plaquette. If the soliton is charged $\pm e$, the phase of the wave function is unchanged; however, if the soliton is neutral, the change in the sign of the wave function is $(-1)^N$, where N is the number of moves of the type shown in (a). (The final move, which takes two horizontal bonds to two vertical bonds, must be done carefully to avoid a degeneracy in the many-body ground state. When this is done, it yields no phase change.) The neutral solitons therefore behave like particles with a negative hopping matrix element. This sign change can also be written (for a loop) as $(-1)^{N'}$ where N' is the number of enclosed plaquettes, as if the neutral soliton were in an external gauge field with half a flux quantum per plaquette. (c) In this path, we exchange two red solitons in such a way that the bonds return to their initial configuration. One soliton moves from a to b along the top path, and the other moves from b to a along the bottom. If the solitons are charged, the wave function is unchanged; if the solitons are neutral, the wave function changes sign. (This path encloses an even number of plaquettes; as noted above, neutral solitons following paths enclosing an odd number of plaquettes will pick up an extra factor of -1.)

ACKNOWLEDGEMENTS

We thank Phil Anderson for inspiration and guidance in this project. We gratefully acknowledge useful discussions with Doug Scalapino, Al Sievers, Bob Buhrman, Shahab Etemad, Peter Littlewood, and Dung-Hai Lee. This work was funded in part by the National Science Foundation under Grants No. DMR-83-18051 (S.A.K.), No. DMR-85-03544 (D.S.R. and J.P.S.), and No. PHY-82-17853. Two of us (S.A.K. and J.P.S.) acknowledge support through the Alfred P. Sloan Foundation.

REFERENCES

1. P. W. Anderson, *Science* , **235**, 1196 (1987); see also L. Pauling, *The Nature of the Chemical Bond*, Cornell Univ. Press, Ithaca, NY, 1960.
2. J. G. Bednorz and K. A. Müller, *Z. Phys. B*, **64**, 188 (1986); S. Uchida, H. Takagi, K. Kawasaki, and S. Tanaka, *Jpn. J. Appl. Phys. Lett.*, to be published; R. J. Cava, R. B. van Dover, B. Batlogg, and E. A. Rietman, *Phys. Rev. Lett.*, **58**, 408 (1987); C. W. Chu, P. H. Hor, R. L. Meng, L. Gao, Z. J. Huang, and Y. Q. Wang, *ibid.*, **58**, 405 (1987).
3. These topological excitations certainly do not exhaust the degrees of freedom. In particular, there will be neutral, transverse phononlike excitations of the RVB degrees of freedom.
4. C. N. Yang, *Rev. Mod. Phys.*, **34**, 694 (1962).
5. V. B. Grande, Hk. Müller-Buschbaum, and M. Schweizer, *Z. Anorg Allg. Chem.*, **48**, 120 (1977); J. M. Longo and P. M. Raccah, *J. Solid State Chem.*, **6**, 526 (1973).
6. Peter Littlewood, private communication.
7. P. W. Anderson, *Mater. Res. Bull*, **8**, 153 (1973).
8. P. Fazekas and P. W. Anderson, *Philos. Mag.*, **30**, 432 (1974).
9. W. P. Su, J. R. Schrieffer, and A. J. Heeger, *Phys. Rev. B*, **22**, 2099 (1980); A. J. Heeger, S. Kivelson, J. R. Schrieffer, and W. P. Su, unpublished.
10. G. Baskaran, private communication; S. A. Kivelson, D. S. Rokhsar, and J. P. Sethna, unpublished.
11. The true RVB ground state will certainly contain an admixture of virtual soliton-antisoliton pairs. However, topological long-range order, similar to that in the XY model, may survive.
12. M. V. Berry, *R. Soc. London, Ser. A*, **392**, 45 (1984); D. P. Arovas, J. R. Schrieffer, and F. Wilczek, *Phys. Rev. Lett.*, **53**, 722 (1984).
13. J. P. Sethna and S. Kivelson, *Phys. Rev. B*, **26**, 3513 (1982); A. Auerbach, and S. Kivelson, *ibid.*, **33**, 8171 (1986).
14. We are assuming that the dispersion relation is not completely changed by the exotic order of the RVB state. Also, our analysis is substantially unchanged in discussing three-dimensional cubic systems, except that the mass

of the solitons will be larger. The binding energy of the charged soliton is not strongly dependent on dimension; thus the possibility exists that bosons could exist above their condensation temperature in three dimensions.

15. L. F. Mattheiss, *Phys. Rev. Lett.*, **58**, 1028 (1987).

16. Both interlayer couplings and attractive interactions between charge solitons (mediated, for example, by exchange of neutral solitons) should allow Bose condensation, which, strictly speaking, will not occur in a noninteracting two-dimensional system.

Z. Zou and P. W. Anderson
Joseph Henry Laboratory of Physics
Jadwin Hall, Princeton University
Princeton, New Jersey 08544

Neutral Fermion, Charge e Boson Excitations in the RVB State and Superconductivity in La$_2$ CuO$_4$-Based Compounds†

We derive the neutral fermion excitation spectrum in the RVB state based on a fermion-boson field theory. It is formally shown that charge e bosons are created upon doping while the neutral fermion excitation remains gapless, predicting a linear low temperature specific heat $C \propto \gamma T$ in the superconducting state. Bose condensation in doped samples leads to the observed superconductivity with true "ODLRO."

Since the original work by Anderson[1] on the resonating valence bond (RVB) theory of high-T_c superconductivity, considerable progress has been made. Baskaran, Zou, and Anderson[2] (BZA) have developed a mean field theory in which many of Anderson's conjectures are confirmed, especially the existence of a pseudo Fermi surface in the insulator. Kivelson, Rokhsar and Sethna[3] have pointed out that there exist three kinds of excitations of RVB states: Fermion solitons, which we call spinons; charge $\pm e$ bosons; and true electrons or holes.

In a recent letter,[4] we identified the mysterious high-T "twitch" transition[5] in La$_2$CuO$_4$ with the mean field RVB transition of the Heisenberg model of BZA; we argued that doping the pure La$_2$CuO$_4$ in the RVB state is compensated by

†Reprinted from *Phys. Rev. B* (in press) (1988)

creation of boson excitations, as long as the system remains in the RVB state; we also argued that after projection with the Mott-Hubbard condition $n_i = 1$, the bare fermions $C_{i\sigma}^+$ turn into "spinon" degrees of freedom $S_{i\sigma}^+$ which have no true kinetic energy in the insulating state; the $S_{i\sigma}$ is treated as strictly neutral, even after doping ($n_i \neq 1$), and spinon excitations remain gapless. This is strongly supported by the linear temperature-dependent low temperature specific heat in the superconducting phase as well as in the normal phase for both La-based and Y-based compounds.[6] In this letter we demonstrate how this is formally done in the Hubbard model with the help of the so called "slave boson" technique developed in the study of heavy fermion systems by Coleman,[7] Barnes[8] etc. The "slave boson" technique was used by Kotliar and Ruckenstein[9] to study the metal-insulator transition for the finite-U Hubbard model, where they derived the same result as that of the Gutzwiller approximation. However, in the previous work the "slave boson" was introduced merely for mathematical convenience (as is implied by the name "slave"). To the best of our knowledge, the physical meaning of the "slave boson" is not well understood. In the present work, we show that in the RVB state "slave bosons" are not slave any more. They have real physical meaning if properly treated; they are related to the soliton holes that are introduced by doping the insulating RVB vacuum state; and they carry a conserved quantum number, charge. A similar but different approach has been used recently by Newns[10] to study the high temperature superconductivity.

We shall start from the finite U Hubbard model:

$$H = -t \sum_{\langle ij \rangle \sigma} C_{i\sigma}^+ C_{j\sigma} + U \sum_i n_{i\uparrow} n_{i\downarrow} - \mu \sum_{i\sigma} C_{i\sigma}^+ C_{i\sigma} \qquad (1)$$

Let's consider site "i": there are four possible states, $|0\rangle$, $|\alpha\rangle$, $|\beta\rangle$ and $|\alpha\beta\rangle$, corresponding to an empty site, one up (down) spin state and a doubly occupied site. Using Hubbard[11] projection operators, we have the completeness relation for each site i:

$$|0\rangle\langle 0| + |\alpha\rangle\langle\alpha| + |\beta\rangle\langle\beta| + |\alpha\beta\rangle\langle\alpha\beta| = 1 \qquad (2)$$

Since $|ip\rangle (p = 0, \alpha, \beta, \alpha\beta)$ form a complete set for site i, any operator affecting only the electron of site i can be written in terms of $|ip\rangle\langle ip|$.[11] In particular

$$C_{i\sigma} = |0\rangle\langle\sigma| + \text{sgn}(\sigma)|-\sigma\rangle\langle\alpha\beta| \qquad (3)$$

where $\sigma = \alpha, \beta$ and $\text{sgn}(\sigma) = +(\sigma = \alpha)$ or $-(\sigma = \beta)$. Note that for $\sigma = \beta$ a "$-$" sign is needed in Eq. (3) to preserve the anticommutation relations.[12]

As was shown by Hubbard,[11] some of the projection operators are fermion-like, some boson-like; their commutators or anticommutators form an algebra. These algebraic properties of projection operators are exactly reproduced by a combined fermion-boson field theory, namely by mapping, for instance, $|0\rangle\langle 0| \rightarrow e_i^+ e_i$, $|0\rangle\langle\alpha| \rightarrow e_i^+ S_{i\alpha}$, $|\alpha\beta\rangle\langle\alpha| \rightarrow d_i^+ S_{i\alpha}$, and $|\alpha\beta\rangle\langle\alpha\beta| \rightarrow d_i^+ d_i$, etc. Here e_i, d_i are boson fields satisfying $[e_i, e_j^+] = \delta_{ij}$, $[d_i, d_j^+] = \delta_{ij}$ and $[e_i, d_i^+] = 0$ etc., and $S_{i\sigma}$ are fermions satisfying $[S_{i\sigma}, S_{j\sigma'}^+]_+ = \delta_{ij}\delta_{\sigma\sigma'}$. One can easily convince oneself that the mapping from projection operators to fermion-boson field theory is self-consistent. Thus the true (bare) electron operators are expressed as, using Eq. (3),

$$C_{i\sigma}^{+} = e_i S_{i\sigma}^{+} + \text{sgn}(\sigma) d_i^{+} S_{i-\sigma} \tag{4}$$

We note that $C_{i\sigma}^{+}$ in Eq. (4) still satisfies the anticommutation relations provided that

$$e_i^{+} e_i + d_i^{+} d_i + \sum_{\sigma} S_{i\sigma}^{+} S_{i\sigma} = 1 \tag{5}$$

corresponding to the completeness conditions (2).

Substituting Eq. (4) into Eq. (1), we obtain

$$H = H_0 + tH' \tag{6}$$

$$H_0 = -t \sum_{\langle ij \rangle \sigma} (e_i e_j^{+} - d_i d_j^{+}) S_{i\sigma}^{+} S_{j\sigma} + U \sum_i d_i^{+} d_i + \mu \sum_i (e_i^{+} e_i - d_i^{+} d_i) - \mu N$$

$$H' = -\sum_{\langle ij \rangle} \left[(e_i d_j + e_j d_i) S_{i\alpha}^{+} S_{j\beta}^{+} + \text{h.c.} \right]$$

where N is the number of the lattice sites. Let's pause for a moment to discuss the nature of the transformation (4). Physically e^{+} corresponds to creating an empty site and d^{+} a doubly occupied site. Therefore e^{+} and d^{+} have opposite charges (e and $-e$, respectively). So we see immediately from Eq. (4) that we can treat $S_{i\sigma}^{+}$ as a neutral particle. One may check the consistency of this assignment by a direct calculation of the currents due to bosons: The current of particle l ($l = e, d,$ or S) is given by

$$\vec{j_l} = \frac{\partial \vec{P_l}}{\partial t} = i[H, \vec{P_l}] \tag{7}$$

where $\vec{P_l}$ is the polarization operator for particle l; in the tight binding model, $\vec{P_l}$ is given by, for instance, $\vec{P_e} = q_e \sum_i \vec{R_i} e_i^{+} e_i$ (with $q_e = e$). A straightforward calculation using the Hamiltonian (6) shows that the total current $\vec{j_c} = \vec{j_e} + \vec{j_d}$, implying that the charges of $S_{i\sigma}$ are identically zero. Thus we can assign the charges of the bare electrons to bosons and the spins to spinons.

One is tempted to do a mean field theory on the Hamiltonian (6) (namely replace e_i and d_i by their classical values).[10] However this is incorrect because the Hamiltonian (6) contains many virtual processes which ought to be eliminated before we let the bosons condense to a zero momentum state. This is achieved by a canonical transformation[13] $\tilde{S}$, which eliminates H' term in H to the order of $t/U : tH' + [H_0, \tilde{S}] = 0$. For sufficiently large U (larger than the critical U_c for the Mott-transition), we neglect terms of order of $(t/U)^2$ and restrict ourselves within the subspace determined by Eq. (5), yielding

$$\tilde{S} = \frac{t}{U} \sum_{\langle ij \rangle} (e_i^{+} d_j^{+} + e_j^{+} d_i^{+}) S_{j\beta} S_{i\alpha} - \text{h.c.}$$

$$H_{\text{eff}} = H_0 - J \sum_{\langle ij \rangle} [S_{i\alpha}^{+} S_{j\beta}^{+} S_{j\beta} S_{i\alpha} + S_{i\alpha}^{+} S_{j\alpha} S_{j\beta}^{+} S_{i\beta}] \tag{8}$$

where $J = 4t^2/U$ and H_0 is defined in (6). One can set $d = 0$ in (8) since the excitations produced by $d^\dagger$ involve large energy U. (Self-consistent solution always leads to $d = 0$ for large U.) One could in principle integrate out fermion degrees of freedom obtaining a free energy functional F; it is easy to see from Eq. (8) that

$$\frac{\partial F}{\partial \mu} = e^2 N - N = -N_0 \qquad (9)$$

where N_0 is the total number of electrons, yielding $e^2 = \delta$ (concentration of holes). Thus we have proved that the boson amplitudes correspond to the hole amplitudes. At low temperature bosons will undergo Bose condensation with true "ODLRO" leading to superconductivity. As we shall see in the following, the Hamiltonian (8) contains essentially almost all the physics we need.

We first discuss the insulating state. The second term in H can be written as $J/4 \sum_{\langle ij \rangle} (\vec{\sigma}_i \cdot \vec{\sigma}_j - 1)$ in the half-filled band, where $\vec{\sigma}/2$ is the spin operator for spinons. There are no bosons in the insulator, $e^2 = d^2 = 0$. Our main purpose in this paper is to calculate the excitation spectrum of spinons and we do not attempt to prove the existence of the RVB state. It is assumed, instead, that we have a RVB ground state[1,4] at $T = 0$, of which a typical configuration of valence bonds is schematically shown in Fig. 1 with all the electrons in the singlet pairs. We want to create excitations on this RVB vacuum state. The simplest excitation one can imagine is a dangling spin[3](spin soliton). It is important to recognize that spin-up and spin-down solitons are always created in pairs; they behave like particle-antiparticle pairs, or in other words the number of spin solitons is not conserved. In the absence of holes, although there are no direct hopping matrix elements for spinons, they can gain coherence energy by exchanging particle pairs with the RVB background. As is illustrated in Fig. 1, spinon and anti-spinon are simultaneously created and then annihilated at the nearest sites, resulting in an equivalent hopping. Based on these physical considerations we adopt the following strategy to calculate the spinon spectrum: (1) We work in a grand canonical ensemble allowing the spinon number to fluctuate between $|N\rangle$ and $|N \pm 2\rangle$, etc., since the kinetic energy of spinons comes entirely from this kind of coherent fluctuations. (2) We then perform a mean field theory on Hamiltonian (8), looking for $\langle S_{i\sigma}^+ S_{j-\sigma}^+ - S_{i-\sigma}^+ S_{j\sigma}^+ \rangle = \Delta$. This has been done in BZA's paper,[2] where the excitation spectrum is given by

$$E_k = \Delta J |\cos k_x + \cos k_y| \qquad (10)$$

In doped systems, the motion of boson holes is compensated by the motion of spinons. The same type of mean field calculation gives the excitation spectrum

$$E_k = E_0 |\cos k_x + \cos k_y| \qquad (11)$$

where $E_0 = (\Delta^2 J^2 + 4t^2 \delta^2)^{1/2}$. The result obtained here is different from that of BZA's paper, where a gap is opened up once we move away from the half filling. The

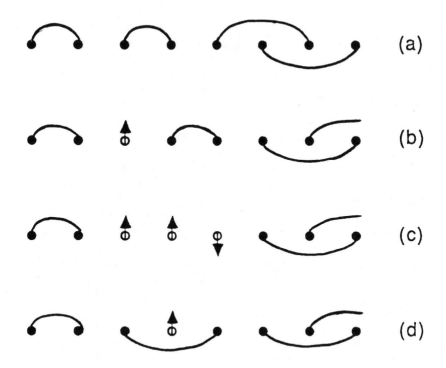

FIGURE 1 (a) A typical configuration of valence bonds in the RVB; (b) to (d), hopping process of a dangling spin (spin soliton) via simultaneous creation and then destruction of a soliton-antisoliton pair in the RVB.

reason that the spinon excitation remains gapless lies in the fact that the number of spinons is not conserved and that the RVB vacuum serves as a reservoir of particles. The chemical potential μ does not appear in the expression for E_k in (11), since the charges are compensated by creation of charge e boson holes (see Eq. (9)). The gaplessness is supported by a theorem due to Lieb and Mattis[14] which applies here without modification. This is an important result because it implies that in the superconducting phase the contribution to specific heat from spinons is linear in temperature $C \propto \gamma T$, with a more or less conventional metallic γ. This prediction is confirmed by experimental data.[6] To compare with the experimental data, we make a rough numerical estimate of γ: effective mass of the spinons is given by $\hbar/E_0 a^2$; a simple exercise of two-dimensional Fermi statistics yields $\gamma \approx \pi k_B^2/6E_0 a^2$. If we take $J = 1000\,\mathrm{K}$ (a reasonable value obtained elsewhere;[15] see also Ref. (4)) and $\delta = 0$ (insulator) we obtain $\gamma \sim 1.8\,\mathrm{mJ/K^2}$· mole, which is close to the observed values. Considering the uncertainty in the parameters (J and t), one should not take this number literally; the actual value may differ by 2 to 3mJ/K^2· mole. Many groups have reported this $C \sim \gamma T$ behavior both in the insulating and superconducting state with slightly different γ. The "variation" of γ can be due to magnetic impurities in the samples, since the spinons carry spins and may be bonded to the impurities, *i.e.*, some of the spinons may be frozen out.

An objection might be raised with regard to the treatment of the constraint (5). The constraint should be strictly enforced in the insulator in order to arrive at the RVB liquid state without breaking the gauge symmetry. However, as we said, we are not deriving the RVB ground state in this paper, but intend mainly to study the excitations of the assumed ground state and try to look at their experimental consequences. The concept of the pseudo Fermi surface (PFS) is useful only for low energy excitations at low T. At low T we have relatively few spin solitons and the particle-antiparticle condition implies $S_{k\sigma}^+ = S_{-k,-\sigma}$ for $\vec{k}$ near the PFS. We can satisfy the constraint by overcompleteness of $S_{i\sigma}$, since one half as many degrees of freedom satisfy the constraint by anticommutation relations. Hence the quasiparticles are no longer subject to the local constraints because they can be annihilated into RVB background or created from the background. A spinon cannot hop directly from site to site unless there are holes next to it. The situation resembles the formation of the heavy fermion band in Kondo lattices in which the localized electrons with opposite spins on neighboring sites have AFM interactions and they can simultaneously hybridize with the conduction electron background[16] forming a coherent heavy fermion band. So we believe that our calculation of the spinon excitation is not bad after all, though more refined work is clearly needed.

Next, we discuss AFM *vs.* RVB. Clearly there is a competition between the AFM and the RVB in the insulator. While we are unable to specify in the present work which state will be energetically more favorable, the AFM state has been observed[17] in "pure" (slightly oxygen-deficient) La$_2$CuO$_4$, therefore the AFM may or may not be more stable than RVB in the insulator. However, as we shall see, a small percentage of doping will destroy this AFM ordering. In the AFM state we have two sublattices "red" and "black" and there are no direct matrix elements connecting the sites on the same sublattice. The hopping term in Eq. (8) vanishes in the AFM because the motion of a hole will induce a string of defects in the ordered state. Thus a large amount of kinetic energy is lost. By balancing the gain in kinetic energy against the energy difference between RVB and AFM, one can show that less than 2% doping will kill the AFM ordering.[4]

The Hamiltonian (8) also describes the normal state. One important conclusion drawn from (8) is that the boson mass is of order of the band electron mass.[4] The hopping term in (8) may be well approximated as

$$-t\rho_0 \sum_k e_k^+ e_k - t \sum_{k \neq k', q\sigma\tau} e^{i(\vec{k}+\vec{k}'-\vec{q})\cdot\vec{r}} e_k^+ e_{k'} S_{-k+q,\sigma}^+ S_{-k'+q,\sigma} \qquad (12)$$

where ρ_0 is the fermion background amplitude. The first term shows that the band width of the bosons is of order t, from which the high T_c results; and the second term describes the scattering processes between holes and spinons, which are of the same order as the band width. In the normal state ($T > T_c$) only those spinons that lie within a range of order T from the PFS are available to scatter the boson holes, therefore we expect a large linear temperature dependent resistivity[4] $\rho \sim T$.

This $\rho \sim T$ behavior is one of the most striking properties of all high-T_c material. A strong experimental fact which may confirm the presence of bosons is the absence of an extrapolated residual resistivity for reasonably good samples. Detailed calculations of $\rho \sim T$ behavior will be published later.

Finally, we address the question of the flux quantization within our formalism: As was discussed by Anderson etc.,[4] the Josephson frequency will still be $2\,eV/\hbar$ in spite of boson charge e. Since the bosons are spinless, it is necessary to require a two-electron tunneling process in order to conserve the spin angular momentum. We also argued in the previous work that holes can have only "ODLRO" and not true macroscopic coherence.[4] We can see this clearly from the following argument based on the present theory: Our Hamiltonian (8) has full $U(1)$ symmetry; it is invariant, in particular, under the transformation P, $PC_iP^{-1} = -C_i$, $Pe_iP^{-1} = -e_i$ and $PS_iP^{-1} = -S_i$. For a fermion system the symmetry P can not be spontaneously broken (Yang's theorem[18]), thus $\langle\Psi|e|\Psi\rangle = 0$. Another way to see this is: if $\langle\Psi|e|\Psi\rangle \neq 0$, we would have a ground state wavefunction Ψ

$$|\Psi\rangle = \sum_N a_N |N, e\rangle \tag{13}$$

where $|N, e\rangle$ is a state with N bosons (N = positive integers). However, we also have the local constraint (5), $e_i^+ e_i + \sum_\sigma C_{i\sigma}^+ C_{i\sigma} = 1$ (since $d_i^+ d_i = 0$). Thus the local fluctuations of the boson number are compensated by electron number fluctuations, leading to

$$|\Psi\rangle = \sum_N \tilde{a}_N |N, C\rangle \tag{14}$$

where $|N, C\rangle$ is a state with N electrons. But an electron system is not allowed to have particle number fluctuations such as $|N\rangle \rightarrow |N \pm 1\rangle$, so we conclude that if a boson is created in one region of the sample, then a boson must be annihilated in another region of the sample. In other words, we have a true 'ODLRO." We therefore believe that the unit of flux quantization in a whole sample is $\hbar c/2e$ not $\hbar c/e$. The situation here differs from that of He^3 and He^4 mixture in which the numbers of He^3 and He^4 atoms can fluctuate independently.

In conclusion, we have derived in this letter the excitation spectrum for the spinon based on a boson-fermion field theory. It is formally shown that the hole due to doping behaves like a boson with band width of order $\sim t$. In the normal state boson-spinon scattering leads to the observed temperature dependence of resistivity $\rho \sim T$; Bose condensation of holes results in the transition to the superconducting state with true "ODLRO," in which the spinon excitation remain gapless giving rise to a linear temperature dependent specific heat. While our theory contains essentially all relevant physics, many questions remain: for example, in the large doping limit the RVB state is destroyed and the holes tend to bind together with the spinons; we expect a crossover from bose-condensed superconducting state to a BCS-like state with order parameter $\langle C_{k\uparrow}^+ C_{-k\downarrow}^+ \rangle$; the cross-over behavior is not well understood at the present. Nonetheless, we believe that the formalism developed here serves as a starting point for further investigations.

ACKNOWLEDGEMENTS

One of us (Z. Z.) is indebted to B. Doucot and P. Coleman for stimulating discussions at the beginning of this work. Zou is grateful to E. Abrahams and G. Baskaran for encouragement during this work and for reviewing the manuscript. Zou also benefited much from numerous discussions with R. Kan, X. Wen and J. Wheatley. We wish to thank T. Hsu, S. Liang, P. Ong, J. Sauls, S. Coppersmith and the entire condensed matter group at Princeton for useful discussions. We also thank H. R. Ott and N. Phillips for early communication of specific heat data and T. H. Geballe for many discussions of experimental data. This work is supported in part by the NSF Grant No. DMR-8518163.

REFERENCES

1. P. W. Anderson, *Science*, V**235**, 1196 (1987).
2. G. Baskaran, Z. Zou and P. W. Anderson, *Solid State Commun.*, **63**, 973 (1987) and this volume; also see A. Ruckenstein, P. Hirschfeld, and J. Appel, *Phys. Rev. B*, **36**, 857 (1987) and this volume; G. Kotliar (1987) preprint.
3. S. Kivelson, D. Rokhsar and J. Sethna, *Phys. Rev. B*, **35**, 8865 (1987) and this volume, and private communications.
4. P. W. Anderson, G. Baskaran, Z. Zou and T. Hsu, *Phys. Rev. Lett.*, **58**, 2790 (1987).
5. R. M. Fleming, *et al.*, *Phys. Rev. B*, **35**, 7191 (1987); T. Fujita, *et al.*, *J.J.A.P.* **26**, 202 (1987).
6. H. R. Ott, private communication; N. Phillips, preprint and private communication; J. E. Crow, private communication; see also the data of a preprint by, M. E. Reeves, T. A. Friedmann and D. M. Ginsberg.
7. P. Coleman *Phys. Rev B*, **29**, 3035 (1984).
8. S. E. Barnes, *J. Phys.*, F**6**, 1375 (1976); and F**7**, 2637 (1977).
9. G. B. Kotliar and A. Ruckenstein, *Phys. Rev. Lett.*, **57**, 1362 (1986).
10. D. M. Newns, preprints and private communication.
11. J. Hubbard, *Proc. R. Soc. A*, **285**, 542 (1965).
12. One of us (Zou) is indebted to B. Doucot for discussions on this point.
13. R. C. Gros, R. Joynt and T. M. Rice, preprint; J. E. Hirsch, *Phys. Rev. Lett.*, **54**, 1317 (1985); also see Ref. 2.
14. E. Lieb and D. Mattis, *J. Math. Phys.*, **3**, 749 (1962); Ian Affleck, private communication.
15. P. W. Anderson, in *Proceedings of The International School of Physics* "Enrico Fermi," July 1987, "Frontiers and Borderlines in Many Particle Physics."
16. M. Noga, preprint.
17. Y. Yamaguchi, *et al.*, *J.J.A.P. Lett.*, to appear; R. Greene, *et al.*, IBM preprint.

18. C. N. Yang, *Rev. Mod. Phys.*, **34**, 694 (1962); we are grateful to E. Abrahams and G. Baskaran for valuable discussions on this point.

P. W. Anderson and Z. Zou
Joseph Henry Laboratory of Physics
Jadwin Hall, Princeton University
Princeton, New Jersey 08544

"Normal" Tunneling and "Normal" Transport: Diagnostics for the RVB State[†]

The "Normal" Transport properties of the high T_c superconductors, especially tunneling and anisotropic resistivity, , are extraordinarily anomalous. We show that these properties can be explained, perhaps uniquely, by a two-dimensional RVB state.

The newly discovered high-T_c oxides exhibit many very strange properties. While the high-T_c itself has generated tremendous excitement, the normal state behaviors are also fascinating. The "normal" transport properties, among other things, are extraordinarily anomalous. It seems to us that much of the theoretical work on high-T_c superconductivity has ignored many of these experimental facts and concentrated only on the high-T_c itself, while the overwhelming probability is that all of them have the same cause.

The RVB theory proposed first by Anderson and subsequently developed by other people,[1] on the other hand, appears to be able to account for most of the experimental observations. The essence of this theory is that the strong electron-electron correlation results in the separation of charge degrees of freedom from spin degrees of freedom. The low energy excitations consist of the charged bosons and

[†]Reprinted from *Phys. Rev. Lett.*, **60**, 132 (1988).

neutral fermions (spinons) with a pseudo Fermi surface. This theory based on the two kinds of excitations has many unusual experimental consequences, many of which have been confirmed. In this work we point out that the "normal" transport experiments, especially tunneling and anisotropic resistivity, provide the strongest evidence that there exist boson and spinon excitations in the RVB state. The understanding of these excitations in the normal state is absolutely crucial towards a final theory of the high-T_c superconductivity.

Traditionally, tunneling has been a very effective probe to understand superconductivity. However, in almost all attempts to study tunneling in the cuprate superconductors the so-called "background" conductance at relatively high voltages is not constant but rises sharply with voltage; in fact, to a best approximation linearly: $\sigma_T \propto |V|$. Similar behavior is often found even at quite low voltages (see Fig. 1).[2-4] In neither case is there any indication of breakdown, Zeller-Giaever effects or inelastic phonon-assisted tunneling, and the behavior is so regular and uniform as almost certainly to be intrinsic. At low voltages, it often seems that a surface layer of the material is not superconducting but "normal."

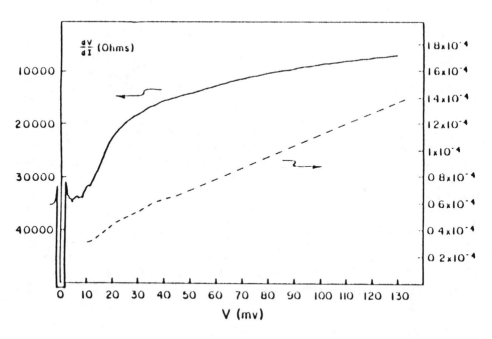

FIGURE 1 A sample of Dynes' tunneling data for single crystal $YBa_2Cu_3O_7$/Pb junction. The upper curve is row date of dV/dI; the straight line is the conductance; the structure at low voltage is due to Pb phonons and the Pb energy gap. The data are taken at $T = 1.4$ K.

Recently Tozer, *et al.*[5] have measured the anisotropic conductivity of a single crystal of $YBa_2Cu_3O_{7-x}$ with very striking results. The unexpected simplicity of the result can be brought out by replotting their data as ρT vs. T^2. In both the *a b* planes and in the *c* direction the resistivity can be fitted very accurately by

$$\rho = \frac{A}{T} + BT \qquad (1)$$

(see Fig. 2) but the coefficients are remarkably different: $A_c = 1.35$, $B_c = 3 \times 10^{-5}$, $A_{ab} = .7 \times 10^{-2}$ and $B_{ab} = 1.4 \times 10^{-6}$. (The resistivity is measured in $\Omega-cm$). We note that a contact misalignment of $1/4$ could account for A_{ab}, and hypothesize that

$$\rho_{ab} = 1.4 \times 10^{-6} T \Omega - cm. \qquad (2)$$

Equally, we suspect that patchiness and defects (and the difficulty of making a 4-terminal measurement) account for the B_c term and—with less certainty— hypothesize that

$$\rho_c = \frac{1.35}{T} \Omega - cm \qquad (3)$$

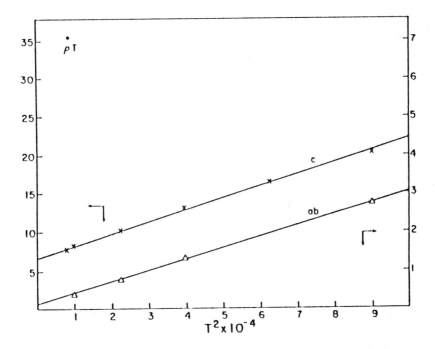

FIGURE 2 A plot of the resistivity data of Tozer, *et al.*: ρT vs. T^2. Note that the scale for ρ_{ab} is different from that of ρ_c.

We have plotted a wide variety of data on ceramic samples both of La-Sr$_2$CuO$_4$ and of "1 2 3" and find that the expression (1) fits many of them very well. Recently Hagen, *et al.*,[14] have tested Eq. (1) in a large number of single crystals of YBa$_2$Cu$_3$O$_7$ between room temperature and T_c and confirmed it very accurately.

We suppose that in fact the materials are all metals in the Cu-O planes and "semiconductors" for conduction between planes and across grain boundaries, in the sense that this conduction is invariably a tunneling process.

All of these observations are compatible with the idea that conduction in the "normal" state is mediated by hole solitons ("kivelsons"[6]) which can be taken to be charged, spinless bosons, and that the magnetic fluctuations in the "normal" metal are Fermion solitons ("spinons") with a pseudo Fermi surface, no charge and spin 1/2. Both of these excitations are solitons of the two-dimensional Cu-O layers, involving rearrangements of the entire layer wave functions, and as such cannot tunnel from one layer to another. The only three-dimensional objects are real electrons, which can tunnel between layers but must then break up into soliton-spinon pairs of excitations.

The charged carriers are boson hole solitons of charge e and they are scattered very effectively by the spinons, of which there is a number $\propto T/J$. Other scattering processes are minor if perhaps not negligible in very poor samples: bosons of relatively long wavelength might have rather long mean free paths for impurity scattering.

Let us derive the above results from the Zou-Anderson-Wheatley theory. This theory leads to an effective "in-plane" Hamiltonian

$$\mathbf{H} = t_{\text{eff}} \sum_{\langle ij \rangle} e_i^+ e_j + t \sum_{\langle ij \rangle} \left(s_{i\sigma} s_{j\sigma}^+ - \langle s_{i\sigma} s_{j\sigma}^+ \rangle \right) e_i^+ e_j + \sum_k \Gamma_k s_{k\sigma}^+ s_{k\sigma} \qquad (4)$$

t_{eff} and $\Gamma_{ij} \sim J$ are renormalized parameters embodying the average effects of the background of singlet pairs. (4) is written in such a way as to bring out the fact that only excitations in the spinon Fermi gas scatter the boson holes, but that this scattering matrix element is as big as the fundamental kinetic energies and is only weakly k-dependent. The spinons are two-fold overcomplete and one may chose to ignore one or the other spin or simply to set $s_{i\sigma} = s_{i-\sigma}^+$.[7]

First we calculate the in-plane conductivity. Roughly this is given by

$$\sigma = \frac{ne^2\tau}{m_B} \qquad (5)$$

and τ in turn can be estimated as the mean scattering free time, neglecting momentum dependence (since with the large Fermi surface Umklapp processes are as common as direct ones). The elastic scattering rate τ^{-1} contains one less power of T for bosons scattered by fermions than for fermions scattering themselves, which is well-known to be $\propto T^2$. This is because there is no exclusion principle restriction on the final state of the boson.

A rough magnitude of the elastic scattering rate is estimated as follows:

$$\frac{1}{\tau} = \frac{2\pi}{\hbar}|t|^2 g_s g_b \int d\xi_q n_F(\xi_q)[1 - n_F(\xi_q)] \tag{6}$$

where g_s is the spinon density of states approximately given[1,7] by $g_s \approx (4\Delta J)^{-1}$, $g_b \approx (4t)^{-1}$ is the boson density of states, n_F is the usual Fermi function and t is the scattering matrix element[7] between holons and spinons. The spinon bandwidth estimated from low-temperature specific heat measurement is of order $(1-2\,\text{eV})^{-1}$. We see that it is not surprising that the resistivity is still linear at $T \sim 500\,\text{K}$ as shown by Gurvitch and Fiory.[8] Within the experimental temperature range $T \ll$ the spinon bandwidth, the integrand in (6) is just a delta function. Thus we obtained $\tau^{-1} = (2\pi/\hbar)t^2 g_s g_b T$. This expression will reach the Mott resistivity at about $T \sim J$, which is the correct order of magnitude according to Ref.(4) and Ref.(7). Note that this scattering rate implies that the spinons are very severely scattered at all T, the mean free path $l_{\text{spinon}} \sim a/\sqrt{\delta}$ and the spinon gas is very disordered.

From the Drude formula (5), the resistivity is given by

$$\rho_{ab} = \frac{m_B^* \pi(U/t)}{32\hbar\Delta n e^2}T \tag{7}$$

where the relation $J = 4t^2/U$ has been used, m_B^* is the boson effective mass and n is the carrier density. A similar but slightly different result has been derived by Isawa et al.[9] Although our estimate is rather crude, the essential physics is included in (7); more sophisticated calculation will only change the numerical factor slightly. If we take $n = 10^{21}/\text{cm}^3$ obtained from Hall effect measurement, $m_B^* \approx m_e$ and $\Delta = 4/\pi^2$,[1] we obtain $\rho_{ab} = 1.3 \times 10^{-6}\,T(\Omega - \text{cm})$ where temperature T is measured in Kelvin.

The process of tunneling between layers is not formally distinct. In this case a boson is scattered not within the layer but from layer to layer, again with emission or absorption of a pair of spinons, one in each layer. The only thing missing is a factor $|T_\perp/t|^2$ which gives the relative amplitude of scattering into the next layer vs. scattering within the layer, $T_\perp$ being the hopping integral from layer to layer. The process in this case is responsible for the conductance rather than the resistance, and we expect the conductance to be less than the Mott conductance by a factor of order $\sigma_c/\sigma_{\text{Mott}} \simeq (T_\perp/t)^2(T/J)$. This seems to give a value for $|T_\perp/t|^2$ of about 0.1 or so.

The interlayer tunneling process is sketched in Fig. 3(a). The dashed line denotes the spinon propagator, the wiggled line denotes the boson propagator and the solid line denotes the true electron propagator. The tunneling probability for bosons from layer i to layer j P_{ij} is again determined by the number of spinons available for scattering

$$P_{ij} = \frac{2\pi}{\hbar}|T_\perp|^2 g_s g_b \int d\xi_q n_F(\xi_q)[1 - n_F(\xi_q)] = \frac{2\pi}{\hbar}|T_\perp|^2 g_s g_b T. \tag{8}$$

From (8) we estimate the conductivity as

$$\sigma_c = \frac{e^2}{\hbar} \cdot 2\pi |T_\perp|^2 g_s g_b^2 T \left(\frac{c}{ab}\right) \tag{9}$$

where $a \cdot b$ is the area of Cu-Cu square and c is the interlayer distance. If we set $|t/T_\perp|^2 = 10$, $t/U = 0.1$ and $J = 1000\,\mathrm{K}$, we estimate the resistivity perpendicular to the Cu-O plane as follows, measuring T in Kelvin, $\rho_c = 1.7/T(\Omega - \mathrm{cm})$ which is in close agreement with the experimental data. We emphasize that these numbers should not be taken literally, since we do not know the precise values of t, U, m^*, etc., and our estimate is rather crude. Nonetheless, the idea that the explanation of the anisotropic resistivity lies in this simple physical picture is hardly to be doubted. Another interesting transport experiments would be the thermopower measurement for which we also expect the anisotropy. The quantitative temperature dependence of the thermopower and thermal conductivityis presently under investigation, but seems reasonable within our picture.

Finally, we consider the normal tunneling. We have calculated this (see below) and find that the obvious phase-space argument gives the answer: The energy must be partitioned between a boson and a spinon, each with essentially constant density of states, and hence the current is proportional to V^2. Here we give a naive calculation of tunneling conductance between a normal metal and a RVB system connected through a weak junction. The electron states on the RVB side are labeled by $(\vec{p}\sigma)$ and on the normal side $(\vec{k}\sigma)$. We assume that the junction is characterized by a tunneling matrix element T_{kp} which is taken to be independent of momentum. The tunneling Hamiltonian is thus given by

$$H_T = \sum_{kp\sigma} \left(T_{kp} c_{p\sigma}^+ a_{k\sigma} + \mathrm{h.c.}\right) \tag{10}$$

The total Hamiltonian of the system is written as $H = H_R + H_N + H_T$, where the subscript $R(N)$ refers to RVB (Normal). All many-body effects are included in H_R. When applying a voltage, the tunneling current is $I(t) = -e\langle \hat{N}_R(t)/dt\rangle$. The time derivative of $\hat{N}_R$ may be obtained from its equation of motion

$$\frac{d\hat{N}_R(t)}{dt} = \frac{i}{\hbar}\left[H, \hat{N}_R\right] = \frac{i}{\hbar}\sum_{kp\sigma}(T_{kp} a_{k\sigma}^+ c_{p\sigma} - \mathrm{h.c.})$$

Following the standard procedure we can express the tunneling current in terms of the single electron spectrum functions[10] $A_R(k,\epsilon)$ and $A_N(p,\xi_p)$

$$I = 2e \sum_{kp} |T_{kp}|^2 \int_{-\infty}^{\infty} d\epsilon\, A_N(k,\epsilon) A_R(p,\epsilon + eV)\left[n_F(\epsilon) - n_F(\epsilon + eV)\right] \tag{11}$$

We shall neglect any many-body effect in the normal metal and use the free electron spectrum function for A_N: $A_N(k,\epsilon) = 2\pi\delta(\epsilon - \xi_k)$. For the RVB state, A_R

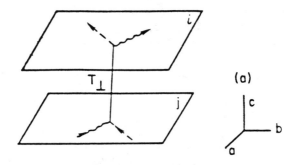

$$\omega_n, q_n = \frac{2\pi(n+1/2)}{\beta}$$

FIGURE 3 (a) A holon grabs a spinon forming an electron, tunnels to the next Cu-O plane and then decays into a holon and spinon; (b) The electron propagator is approximated as a convolution of a holon propagator and a spinon propagator.

is obtained by calculating the imaginary part of the single-particle propagator $G_1(p, \tau - \tau') = -\langle T_\tau c_{p\sigma}(\tau) c_{p\sigma}^+(\tau') \rangle$. G_1 contains complicated many body effects which we were unable to calculate exactly. However in the RVB state, the charge boson excitations are separated from the neutral fermion excitations in the low energy limit. There are no real electron-like quasiparticles present at low temperature $T \ll$ spinon bandwidth. Therefore an electron tunneling into the RVB side from the normal side will have to decay into a boson and a spinon excitation. This process is sketched in Fig. 3(b). To the lowest order approximation, we will replace the spinon and boson propagators by corresponding free particle propagators, respectively. Thus we may approximate G_1 by

$$G_1(p, i\omega_n) = \frac{1}{N\beta} \sum_{q, iq_n} B(q - p, iq_n - i\omega_n) F(q, iq_n) \qquad (12)$$

where $B(F)$ is the free boson (spinon) propagator, N is the number of lattice sites, ω_n and q_n are fermionic Matsubara frequences. The frequency summation is easily carried out and we obtain the spectrum function from (12):

$$A_R(p, \epsilon) = -2Im\, G_{ret}(p, \epsilon)$$
$$= \frac{2\pi}{N} \sum_q \left[n_F(\xi_q) + n_B(\eta_{q-p}) \right] \delta(\epsilon + \eta_{q-p} - \xi_q) \qquad (13)$$

where $\eta_{q-p} = (\vec{q}-\vec{p})^2/2m_B^* - \mu \geq 0$ is the boson energy measured from the chemical potential, which is always non-negative, and n_B is the usual bose function. Note that this spectrum function satisfies the sum rule

$$\frac{1}{2\pi} \int_{-\infty}^{\infty} d\epsilon\, A(p, \epsilon) = 1,$$

as it must. As we can see from (13) that there are no real electron-like excitations at least for $T > T_c$.

Substituting A_N and A_R back into Eq. (11), one can carry out the integral at zero temperature, the final result is

$$I(V) = \frac{2|e|}{\hbar} \pi |T|^2 g_n g_s \left\{ g_B (eV)^2 \left[\theta(V) - \theta(-V) \right] + \delta |e|V \right\}, \qquad (14)$$

where g_n is the density of states of the normal metal. Note that in calculating the integral the condition $\eta \geq 0$ imposes a limit on the integration range. The physical meaning of the second term in (14) is obvious: it describes a tunneling process in which an electron of energy V tunnels into a spinon of the same energy plus a boson of zero energy, namely the electron spin is carried by the spinon and the electron charge is dumped into the "condensate." This term is responsible for the zero-bias conductance as observed by Dynes.[4] From Eq. (14) the linear voltage-dependent tunneling conductance results

$$\sigma_T = \frac{e^2}{\hbar} 4\pi |T|^2 g_n g_s g_B |eV| + \sigma_0, \qquad (15)$$

with the zero-bias conductance σ_0 given by

$$\sigma_0 = \frac{4e^2}{\hbar} \pi |T|^2 g_n g_s\, \delta. \qquad (16)$$

At finite temperature the integral is more complicated, but we do not expect a drastic change in $I(V)$.

We note the fact that the tunneling density of states is a direct measurement of $Im G_1(k, \omega)$,[10] therefore, the effect of superconductivity on G_1 must be quite small, involving its low energy tail only; perhaps tunneling is not a very satisfactory probe of the superconducting properties, as many experimentalists will testify. The big effects are in the collective soliton states, which appear in the density-density correlation function but not in G_1.

Thus the three linear phenomena—tunneling, normal state resistivity parallel to Cu-O plane, normal state conductivity perpendicular to Cu-O plane—rather unequivocally point to the simple two dimensional spinon-holon picture of the normal state above T_c. In this state there are no identifiable electron quasiparticles present within hundreds of degrees of the Fermi energy and it would appear that "conventional" superconductivity is the last thing the system has on its mind, in agreement with simulations.[11]

We suggest that the observed superconductivity must, therefore, be a result of the tunneling between the layers. In a succeeding paper[12] we will show how this tunneling can lead to electron pairing through the boson amplitude and to conventional superconductivity.

In the mean time, what of the electromagnetic properties of our two-dimensional layer? May[13] has proposed that all two-dimensional systems of free bosons are superconductors, but he failed to take into account the Kosterlitz-Thouless process of vortex separation. Any reasonable estimate puts the K-T transition T_c for a system higher than the observed T_c. We are unable as yet to resolve the question of whether the individual layers are independently topologically ordered or not.

ACKNOWLEDGEMENTS

We wish to thank B. Doucot, R. Kan, especially T. Hsu and J. Wheatley for numerous stimulating discussions. We are indebted to R. C. Dynes for allowing us to use the tunneling data prior to publication and N. P. Ong for early communications of the tunneling data. This work was supported by National Science Foundation Grant No. DMR 8518163.

REFERENCES

1. P. W. Anderson, *Science*, **235**, 1196 (1987); G. Baskaran, Z. Zou, and P. W. Anderson, *Sol. St. Comm.*, **63**, 973 (1987) and this volume; S. Kivelson, D. Rokhsar, and J. Sethna, *Phys. Rev. B, Rapid Comm.*, **35**, 8865 (1987) and this volume ; P. W. Anderson, G. Baskaran, Z. Zou and T. Hsu, *Phys. Rev. Lett.*, **58**, 2790 (1987); A. Ruckenstein, P. Hirschfeld, and J. Appel, (1987) *Phys. Rev. B*,**36**,857 (1987) and this volume; Kotliar (1987) MIT preprint.

2. N. P. Ong, private communication.

3. K. E. Gray, M. E. Hawley, and M. R. Moog, in *Novel Superconductivity*, ed. by S. A. Wolf and V. Kresin, Plenum, New York, 1987.

4. R. C. Dynes, private communication; Fig. 1 is a sample of Dyne's data, which clearly shows that the observed behavior is intrinsic.

5. S. W. Tozer, *et al., Phys. Rev. Lett.*, **59**, 1768 (1987).

6. P. W. Anderson, "Frontiers and Borderlines in Many Particle Physics," *International School of Physics* "Enrico Fermi" Course 104, ed. by J. R. Schrieffer and R. A. Broglia, North-Holland, Amsterdam, 1987.

7. Z. Zou and P. W. Anderson, *Phys. Rev. B*, (in press)(1988) and this volume.

8. M. Gurvitch and A. T. Fiory (1987) preprint.

9. Y. Isawa, S. Maekawa and H. Ebisawa in *Proceedings of the Eighteenth Yamada Conference on Superconductivity in Highly Correlated Fermion Systems, Sendai, Japan, 1987*, North-Holland, Amsterdam, 1987.

10. See for instance, J. R. Schrieffer, D. J. Scalapino, and J. W. Wilkins, *Phys. Rev. Lett.*, **10**, 336 (1963).

11. J. Hirsch and Q. Lin, (1987) preprint.

12. J. Wheatley, T. Hsu, and P. W. Anderson, to be published.

13. R. M. May, *Phys. Rev.*, **115**, 254 (1959).

14. S. J. Hagen, J. W. Jing, Z. Z. Wang, J Horvath, and N. P. Ong, to be published.

Robert Joynt
Department of Physics
University of Wisconsin
Madison, Wisconsin 53706

Numerical Studies of Superconductivity in the Two-Dimensional Hubbard Model

The question of superconductivity in the two-dimensional Hubbard model is reviewed. Emphasis is placed on the large-U limit of the single-band model and the relevance of this model to high-temperature superconductivity is discussed. Variational Monte Carlo work has shown that a nominally superconducting Gutzwiller or RVB state has essentially the same energy as the antiferromagnetic state for the half-filled (spin only) case. This state becomes a true d-wave superconductor for the less-than-half-filled (doped) case. This is an interference effect in which the phasing of the Cooper pairs acts to enchance antiferromagnetic spin correlations.

INTRODUCTION

The high-T_c superconductors have hit the newspapers because of their technological promise, but they are also truly fascinating from the point of view of the pure scientist. For one thing, they are very accurately two-dimensional, probably the richest of possible dimensionalities for quantum theory and statistical mechanics, where it seems almost anything can happen especially if there is residual three-dimensional coupling to stabilize exotic ground states at finite temperatures. For another, they are metallic oxides and therefore belong to a class of compounds for which many-body effects are well-known to dominate the physics. We theorists are supping on rich broth these days and it looks pretty certain that it's going to continue for a while. It has already gone on for longer than expected because problems in sample preparations prevented the appetite for new theories from being sated. That time is now coming to an end and cold data which must be confronted is coming out.

The focus of this talk is on a particular model of the oxides, namely the single-band Hubbard model at large U. The next section is devoted to a defense of this model from a fundamental point of view. The third section is devoted to a numerical demonstration that the model does indeed show superconductivity. This answers the objection to the large-U Hubbard picture that that boring old model can't be superconducting anyway so why worry about it. The fourth section contains a comparison of our work to that of others, and some conclusions and discussion.

RELEVANCE OF THE ONE-BAND HUBBARD MODEL

The Hubbard model for a single band of electrons is

$$H = \sum_{ij} t_{ij} c_i^+ c_j + U \sum_i n_{i_\uparrow} n_{i_\downarrow}, \tag{1}$$

where i is a label for the central site of an unspecified Wannier function. The ideas behind this Hamiltonian, particularly as a model for transition metal oxides, are contained in a paper of Anderson,[1] now almost thirty years old. This paper concentrates on compounds with 1-1 metal-oxygen stoichiometry and includes estimates of U. In these compounds the oxygen p-bands are full and U is defined by turning off t and making a local move of an electron between metal sites (and therefore includes also longer range interactions). Because the second and third ionization potential is essentially constant across the transition metal series, so is U and it comes out at 6 to 8 eV. Thus the Wannier functions sit mainly on the cations, t_{ij} has very accurately nearest-neighbor form and U/t is the order 10 or 20. In many cases the crystal symmetry splits the d-band and one is left with a single-band model. These are the canonical large-U systems, to which CuO belongs and of which NiO is the best studied.[2]

The most striking feature of these stoichiometric compounds is that with the antiferromagnetismof stoichiometric oxides band structure calculationsand antiferromagnetism in oxides exception of TiO they are antiferromagnetic insulators. This conflicts with predictions of metallic behavior made by counting electrons and d-bands and forgetting about interactions, but it also disagrees with all band structure calculations which treat interactions as a mean field.[3] This is perhaps not surprising, since the excitations have a local excitonic nature whose energy is excitonic natureof excitations in oxides dominated by relaxation not included in such calculations.

The essence of these local excitations is best captured by cluster calculations,[4] and these are in fact in very good quantitative agreement with spectroscopic data, as demonstrated very convincingly by Sawatzky and Allen[5] for NiO. These authors showed that a puzzling low energy peak (above the Fermi level) in the inverse photoemission intensity was extrinsic, and could identify all the other features with local charge transfers. This analysis contained one big surprise, which is that the excitation of lowest energy (4.3 eV) is a transfer of an electron from an oxygen to a nickel state. The only way to reconcile this with the band calculations, which after all must give the centroids of the p- and d-bands correctly, is to assume that the onsite interaction on the O atoms is larger than expected, or that the site energy is higher.

This brings us up to high-T_c materials. Undoped La_2CuO_4 is the simplest version. The stoichiometric compound is an insulator and an antiferromagnet[7] with formal charges on the Cu and O the same as in CuO. It shares with the other oxides the property of being something of an embarrassment to band theory, which predicts that it is a metal or a charge density wave.[8,9] In these respects it is a typical oxide. In others, however, it is rather different. Most importantly, the p-bands in the band calculations hybridize strongly with the d-bands, so here it is certainly no longer true that we can ignore the p-bands as in Anderson's 1959 picture. In fact, coupled with what we now know about NiO, we might expect that the lowest energy excitations of La_2CuO_4 would involve the creation of O^-. This has been confirmed recently by calculations with frozen electron counts in Cu-O chains,[10] where the U value for the oxygen sites was found to be similar to that on the Cu. These are all very good theoretical reasons to expect that when the compound is doped with Sr or Ba, the resulting holes will sit on the oxygen sites. This is confirmed by all spectroscopic studies[11] which show that Cu^{3+} is absent, (again in conflict with mean-field band theory).

In $YBa_2Cu_3O_7$, the situation is very similar, both experimentally and theoretically. Resonant photoemission has again revealed that Cu^{3+} is nowhere to be found in these materials,[12-14] ruling out the picture of wide, well-hybridized p-d bands. Instead one must go to a picture where the holes sit on oxygen sites. When the hole concentration is small, then there is a narrowing of this O-band due to blocking effects first calculated quantitatively by Gutzwiller. The band is narrowed by a factor $(1-n)$ where n is the number of electrons per unit cell. Recent angle-resolved photoemission work on single-crystal $YBa_2Cu_3O_7$ has demonstrated in detail that this picture is correct.[15] These studies show that the band which crosses the Fermi energy is narrowed as compared to its band-structure-calculated value by a factor of three, in very good agreement with Gutzwiller's formula, when the nominal value $n = 2/3$ is used for the electron density.

The issue is whether, given that the charge degrees of freedom are on the O-sites but there remains a d^9Cu ion which is spin 1/2, we can still speak of a one-band Hubbard model. This question was answered in the affirmative by Anderson recently,[17] and the Wannier functions have been explicitly constructed by Zhang and Rice.[18] In La_2CuO_4, each Cu spin is surrounded by four O atoms. The O-O overlap is negligible. The hole distributed on these four atoms therefore forms a singlet with the Cu spin to gain the large superexchange energy in the cell. This linear combination is well separated from all others in energy. These singlets can then be combined to form Bloch waves and their dispersion can be calculated. The result is a single tight-binding band with nearest-neighbor hopping only, to a good approximation. We are therefore led back to (1). To reiterate, this single band picture *must* hold if there is no Cu^{3+} in the system. All spectroscopic experiments are telling us this.

VARIATIONAL NUMERICAL STUDIES

Once we have decided that (1) is a reasonable model for the electronic system, then the key question is whether superconductivity is lurking in it or not. The first step is to derive an effective Hamiltonian for the large-U limit. For the nearest-neighbor model with small hole concentration, this is simply

$$H_{\text{eff}} = -t \sum_{\langle ij \rangle} (1 - n_{i-\sigma}) c_{i\sigma}^{+} c_{j\sigma} (1 - n_{j-\sigma})$$

$$+ \frac{4t^2}{U} \sum_{\langle ij \rangle} (\vec{S}_i \cdot \vec{S}_j - 1/4). \tag{2}$$

This is just an antiferromagnetic Heisenberg Hamiltonian with a hopping term. The hopping term only moves holes around. It never creates doubly occupied sites. In order of magnitude, the first term is of order $t(1 - n)$ and the second term is of order $t^2 n/U$. the interesting region of parameter space is where these two are of the same order of magnitude.

A good variational Ansatz for the ground state of (2) is

$$|\psi_G\rangle = P_{D=0}|\psi\rangle \tag{3}$$

where $P_{D=0}$ is the projection onto the subspace in which there are no doubly occupied sites and $|\psi\rangle$ is the filled Fermi sea, *i.e.*, the non-interacting ground state. This is the Gutzwiller wavefunction. It is also identical to Anderson's RVB state, as shown by Yokoyama and Shiba.[18] In going from H to H_{eff} we made a canonical transformation which mixed in nearest-neighbor pairs of doubly occupied and hole sites, which means that these virtual fluctuations are included. Real fluctuations, with energy U, are excluded by the projection.

In a half-filled band $|\psi_G\rangle$ represents a Mott insulator. What is remarkable, however, is that $|\psi_G\rangle$ also has strong antiferromagnetic correlations.[19] In one dimension the energy of the wavefunction, which comes from the $\vec{S}_i \cdot \vec{S}_j$ term alone, is within 0.2% of the exact Bethe Ansatz solution for the Heisenberg antiferromagnet.[20] In addition, it has the same long-range power law correlations: $\langle \vec{S}_o \cdot \vec{S}_n \rangle \sim (-1)^n/n$, as in the exact solution.[21,22] In two dimensions on the square lattice, the ground state of the Heisenberg antiferromagnetic Hamiltonian has long been thought to have long range order, with $\langle S_i \cdot S_j \rangle = -.32$ for i, j nearest neighbors. Here the wave function (3) has an energy about 15% higher than this estimate (the exact solution is not known). But if $P_{D=0}|\psi\rangle$ is replaced by $P_{D=0}|\psi_{AF}\rangle$ where $|\psi_{AF}\rangle$ is the Hartree-Fock spin density wave, then the energy is again within 1% of the best values.[18] Note that the naive Néel state is higher in energy than all of these states.

Of course no half-filled band Gutzwiller wavefunction can be superconducting. One can formally write down the BCS wavefunction in first-quantized form in the subspace where the number of particles is fixed to be $n = 1$, and perform the projection:

$$|\psi_G^{BCS}\rangle = P_{D=0} A \left[a(\vec{r}_1 - \vec{r_1}')a(\vec{r}_2 - \vec{r_2}') \ldots \right].$$

Here A is the antisymmetrization operator and $a(\vec{r}_1 - \vec{r_1}')$ is the pair wavefunction for an up spin particle at $\vec{r}_1$ and a down spin particle at $\vec{r_1}'$ in a spin singlet state. Gros[23] showed that this can be written as a determinant:

$$|\psi_g^{BCS}\rangle = \begin{vmatrix} a(\vec{r}_1 - \vec{r_1}') & a(\vec{r}_1 - \vec{r_2}') & a(\vec{r}_1 - \vec{r_3}') & \cdots \\ a(\vec{r}_2 - \vec{r_1}') & a(\vec{r}_2 - \vec{r_2}') & \cdots & \\ a(\vec{r}_3 - \vec{r_1}') & & \ddots & \\ \vdots & & & \end{vmatrix}$$

This means that Monte Carlo methods can be used to evaluate expectation values such as $\langle \psi_G^{BCS}|\vec{S}_i \cdot \vec{S}_j|\psi_G^{BCS}\rangle$. It then turns out in two dimensions that a d-wave "superconducting" wavefunction with $\Delta = \Delta_o(\cos k_x - \cos k_y)$ and $\Delta_o \approx t$ has an energy which is the same, to within the accuracy of Gros' calculations, as the best previous estimates. Of course, this state has no long range order. The energy of s-state superconducting wavefunctions can also be calculated, but have higher energy than the normal state, and are therefore not very interesting.

All of these calculations are for $n = 1$ and do not reflect metallic behavior. One surprising lesson to be learned from them is that a wavefunction which is designed to suppress charge fluctuations with the minimum gain in kinetic energy can also show very strong spin correlations. Another is that introduction of d-wave phasing of the wavefunction can enhance these correlations.

For $n < 1$, the properties of $|\psi_G\rangle$ in one dimension have been extensively investigated by Gros, Joynt, and Rice, again by the Monte Carlo method.[20] The result for the energy is still extremely good, within 10% or so, of the exact solution for all $n < 1$. The kinetic energy is therefore also well handled by (3), as expected from the method of its construction, i.e., building in the filled Fermi sea at the start. This gives hope that the two-dimensional metal will also yield to this Ansatz.

In two dimensions on the square lattice itinerant antiferromagnetism has been investigated by Yokoyama and Shiba[18] using the Gutzwiller wavefunction. They find that the long range order decreases very fast as holes are added to the half-filled band and the ground state reverts to paramagnetism. This was also found by Hirsch using the finite-temperature method.[24]

These papers did not consider superconducting states, but the energy of such states has also been obtained by Gros. He found that a d-wave superconductor, this time a real superconductor, was a locally stable state for a large range of n and t/U. The energy gain by making the superconducting pairing all comes from the spin correlation energy as might be expected from the half-filled case. The condensation energy is reduced because the kinetic energy is increased, as in conventional BCS theory. The optimal value of Δ_o is approximately t for $n = 0.9$ and $U = 16t$ and the condensation energy is then calculated to be $0.01t$. If $t \approx 1\,\text{eV}$ then this is equivalent to a temperature of $100\,\text{K}$, which is a nice round number.

To this point, I have given results but have not stated the physical mechanism which produces the superconductivity. The first thing to say is that it is *not* the obvious mechanism whereby two holes stick together and move together in an anti-ferromagnetic background in order thereby to better preserve a Néel arrangement of spins. This is not the way the pairing works, according to the calculations of Gros and in the Cooper pair computations of Gros, Joynt, and Rice.[25] The correlation function of two holes on nearest-neighbor sites was measured in the superconducting state and shows no enhancement over the normal state. The actual mechanism is much more subtle and is revealed in the Cooper pair results, which first showed that energy is gained by d-wave pairing in the large-U model. If one envisions the Fermi surface of the two-dimensional half-filled tight binding band, it is a square with corners at $(\pm\pi, 0)$ and $(0, \pm\pi)$. Now put an up hole and down hole in these four $\vec{k}$-states, labelled $\vec{k}_1 = (\pi, 0)$, $\vec{k}_2 = (0, \pi)$, $\vec{k}_3 = (-\pi, 0)$ and $\vec{k}_4 = (0, -\pi)$. A normal state is just $|\psi_N\rangle = P_{D=0}c_{k_{i\uparrow}}c_{-k_{i\downarrow}}|\psi\rangle$, where $|\psi\rangle$ is the half-filled band and i is arbitrary. A superconducting wavefunction of the simplest sort is

$$|\psi_{CP}\rangle = P_{D=0}\big(c_{\vec{k}_{1\uparrow}}c_{-\vec{k}_{1\downarrow}} \pm c_{\vec{k}_{2\uparrow}}c_{-\vec{k}_{2\downarrow}} + c_{k_{3\uparrow}}c_{-k_{3\downarrow}} \pm c_{k_{4\uparrow}}c_{-k_{4\downarrow}}\big)|\psi\rangle$$

Here the upper and lower signs correspond to s- and d-wave pairing, respectively. This wavefunction, which has little variational freedom, shows quite clearly that the normal state is unstable. The energy of the d-wave state is lower by about the energy of 1.5 antiferromagnetic bonds per Cooper pair. S-wave pairing raises the energy. The reason is as follows. Consider the overlap $\langle\psi_{\text{Néel}}|\psi_{CP}\rangle \cdot |\psi_{\text{Néel}}\rangle$ is just a checkerboard pattern of fixed up spins $\vec{r}_i$, and downs $\vec{r}_i{}'$. The overlap is a sum of four terms each consisting of an up spin Slater determinant times a down spin Slater determinant. The first up-spin determinant has a first row $e^{i\vec{k}_2 \cdot \vec{r}_1}$, $e^{i\vec{k}_2 \cdot \vec{r}_2}$, etc., while the second has a first row $e^{i\vec{k}_1 \cdot \vec{r}_1}$, $e^{i\vec{k}_1 \cdot \vec{r}_2}$, etc. They are otherwise identical. But note that $\vec{k}_2 - \vec{k}_1 = (\pi, \pi)$ and therefore $e^{i\vec{k}_1 \cdot \vec{r}_i} = e^{i\vec{k}_2 \cdot \vec{r}_i}$, since $\vec{r}_i$ is of the form (x, y) where $x + y = $ even. Similarly $e^{i\vec{k}_1 \cdot \vec{r}_i{}'} = -e^{i\vec{k}_2 \cdot \vec{r}_i{}'}$ since the down spin $\vec{r}_i{}'$ has $x' + y' = $ odd. The maximum overlap is therefore when the d-wave spin signs are chosen for the determinants, since this leads to constructive interference. S-wave gives destructive interference. We see that the resulting correlations are a completely quantum-mechanical effect which cannot be derived from a semiclassical picture of localized holes moving in a Ising-spin background.

DISCUSSIONS AND CONCLUSIONS

The results presented in the last section we obtained by sampling of spatial configurations, and are therefore subject to statistical error. In one dimension, special methods can reduce this error to less than 1%, but in two dimensions, it is generally 5 to 10%. This is small enough to discern all qualitative trends except for one case. That is the question of whether the "superconducting" or antiferromagnetic state is lower in energy at $n = 1$. The near degeneracy of these two phases makes this question difficult to answer. It is therefore not known whether the system is a real superconductor at infinitesimal doping and, if not, what is the critical doping level. More accurate calculations may eventually resolve this important question. The doping level is likely to be rather small, due to the very rapid destruction of the spin ordering by holes on the antiferromagnet.

The computations are done on finite-size samples, with periodic boundary conditions, up to N_s = number of sites = 170. Because of the error thereby introduced it is always necessary to examine a series of samples and extrapolate to the limit of infinite size. Empirically it is nearly always the case that there is a smooth systematic $1\sqrt{N_s}$ dependence in any quantity which, in the $N_s \to \infty$ limits, is intensive. This suggests that the boundary conditions introduce an error which is proportional to the length of the boundary.

Lin and Hirsch[25] have also done calculations on the two-dimensional Hubbard model with the quantum Monte Carlo method. This is a finite-temperature method and the lowest temperatures obtained are $T = 0.08t$ on lattices with $N_s = 36$ and 16. They calculate superconducting susceptibilities and see no sign of any enhancement as T is lowered. It seems safe to say that these calculations are being done in a different regime from the variational ones, since the latter show that the condensation energy is approximately $0.01t$. Extrapolation to infinite size has not yet been attempted with the finite temperature method but might yield interesting results. This may be difficult if $N_s = 36$ is the largest size obtainable, since at this size the majority of sites are on the perimeter.

The chief limitation of the variational approach at the present time is that it has not been extended to excited states. The difficulty is that although it is easy to construct quasiparticle states in the BCS wavefunction, after projection these states are no longer orthogonal, and are vastly overcomplete in the physical subspace. This difficulty has been ignored in thermodynamic mean field theories, which compute entropy counting unphysical states. However, it is not possible to improve on these theories at finite temperature with numerical calculations at the present time. In particular, a numerical calculation of T_c is certainly a long way off.

We can, however, conclude that the antiferromagnetic correlations are reasonably well understood. They arise naturally on the large-U limit of the Hubbard model and do not have to be put in by including more complicated interactions such as, for example, RKKY. This explains the neutron scattering results on both heavy fermion materials[27] and La_2CuO_4.[28] Further, superconductivity is there in the two-dimensional Hubbard model at large U. It would be difficult to believe this without the numerical calculations, since the fundamental interactions are all repulsive. It is therefore remarkable that it was suggested at a very early stage,[29] before the numerical work was done. This work has demonstrated the correctness of the suggestion. It has also shown that d-wave phasing of the pairs produces antiferromagnetic correlations. This is clearly related to earlier work on id-wave superconductivity by exchange of anti-paramagnons[30-32] but it is not known whether the analogy is exact. The calculations have elucidated the symmetry of the phase, and the superconducting mechanism.

ACKNOWLEDGEMENTS

I would like to thank my collaborators T. M. Rice and C. Gros for many discussions on the topics treated here.

REFERENCES

1. P. W. Anderson, *Phys. Rev.*, **115**, 2 (1959).
2. D. Adler, in *Solid State Physics*, Vol. 21, eds. F. Seitz, D. Turnbull, and H. Ehrenreich, Academic Press, New York, 1968.
3. K. Terakura, A. R. Williams, T. Oguchi, and J. Kübler, *Phys. Rev. Lett.*, **52**, 1830 (1984); J. Hugel and C. Carabatos, *J. Phys. C*, **16** 6713 (1983).
4. A. Fujimori and F. Minami, *Phys. Rev. B*, **29**, 5225 (1984), and **30**, 957 (1984).
5. G. A. Sawatzky and J. W. Allen, *Phys. Rev. Lett.*, **53**, 2339 (1984).
6. H. Scheidt, M. Glöbl, and U. Dose, *Surf. Sci*, **112**, 97 (1981); F. J. Himpsel and Th. Fauster, *Phys. Rev. Lett.*, **49**, 1583 (1982).
7. D. Vaknin, S. K. Sinha, D. E. Moncton, D. C. Johnson, J. Newsam, C. R. Safinya, H. E. King, *Phys. Rev. Lett.*, **58**, 2802 (1987).
8. L. F. Mattheiss, *Phys. Rev. Lett.*, **58**, 1028 (1987).
9. J. Yu, A. J. Freeman, and J.-H. Xu, *Phys. Rev. Lett.*, **58**, 1035 (1987).
10. B. Harmon, private communication.
11. A. Fujimori, E. Takayama-Muromachi, Y. Uchida, and B. Okai, *Phys. Rev. B*, **35**, 8814 (1987); J. M. Tranguada, S. M. Heald, A. R. Moodenbaugh, and M.

Suenaga, *Phys. Rev. B*, **35**, 7187 (1987); M. Nücker, J. Fink, B. Reuker, D. Ewert, C. Politis, P. J. W. Weijs, J. C. Fuggle, *Z. Phys. B*, **67**, 9 (1987).

12. M. Onellion, Y. Chang, D. W. Niles, R. Joynt, G. Margaritondo, N. G. Stoffel, and J. M. Tarascon, *Phys. Rev. B.*, **36**, 819 (1987).

13. F. C Brown, T.-C. Chiang, T. A. Friedmann, P. M. Ginsberg, G. N. Kwawer, and T. Miller, and M. G. Mason, *J. Low. Temp. Phys.*, **69**, 151 (1987).

14. R. L. Kurtz, R. L. Stockbauer, D. Mueller, A. Shih, L. E. Toth, M. Osofsky, and S. A. Wolf, *Phys. Rev. B*, **35**, 8818 (1987).

15. N. G. Stoffel, Y. Chang, M. K. Kelly, L. Dottl, M. Onellion, P. A. Morris, W. A. Bonner, and G. Margaritondo, to be published.

16. P. W. Anderson, "Frontiers and Borderlines in Many Particle Physics," *International School of Physics* "Enrico Fermi" Course 104, ed. by J. R. Schrieffer and R. A. Broglia, North-Holland, Amsterdam, 1987.

17. F.-C. Zhang and T. M. Rice, to be published.

18. H. Yokoyama and H. Shiba, to be published.

19. T. A. Kaplan, P. Horsch, and P. Fulde, *Phys. Rev. Lett.*, **49**, 889 (1982).

20. C. Gros, R. Joynt, and T. M. Rice, *Phys. Rev. B*, **36**, 381 (1987).

21. P. Horsch and T. A. Kaplan, *J. Phys.*, C **16**, L1203 (1983).

22. F. Gebhard and D. Vollhardt, *Phys. Rev. Lett.*, **59**, 1472 (1987).

23. C. Gros, to be published.

24. J. E. Hirsch, *Phys. Rev. B.* **31**, 4403 (1985).

25. C. Gros, R. Joynt and T. M. Rice, *Z. Phys. B*, **68**, 425 (1987).

26. H. Lin and J. E. Hirsch, to be published.

27. A. I. Goldman, G. Shirane, G. Aeppli, E. Bucher, and J. Hufnagl, *J. Mag. Mag. Mat.* , **63–64**, 372 (1987).

28. G. Shirane, Y. Endoh, R. J. Birgeneau, M. A. Kastner, Y. Hidaka, M. Oda, M. Suzuki, and T. Murakami, *Phys. Rev. Lett.*, **59**, 1613 (1987).

29. P. W. Anderson, *Science*, **235**, 1196 (1987).

30. V. J. Emery, *J. Phys. Colloq.*, **44**, C3-977 (1983), and *Syn. Metals*, **13**, 21 (1986).

31. D. J. Scalapino, E. Loh, and J. E. Hirsch, *Phys. Rev. B*, **34**, 8190 (1986).

32. K. Miyake, S. Schmitt-Rink, and C. M. Varma, *Phys. Rev. B*, **34**, 6554 (1986).

Ian Affleck
Physics Department
University of British Columbia
Vancouver, BC, V6T 2A6, CANADA

Spin-Gap and Symmetry-Breaking in CuO$_2$ Layers and Other Antiferromagnets[†]

We discuss the possibility of magnetically disordered ground states in anti-ferromagnets and argue that a variety of systems, including the CuO$_2$ layers in the high-T_c superconductors, should either have a spontaneously broken symmetry or else gapless excitations. Models which may have unbroken symmetry and a gap, possibly including the honeycomb lattice $s = 1/2$ antiferromagnet are also discussed.

INTRODUCTION

Lieb, Schultz and Mattis proved[1] a remarkable theorem in 1961 which states that the $s = 1/2$ antiferromagnetic periodic chain of length L has a low energy excitation of O($1/L$). Very recently it was observed[2] that this theorem can be trivially extended to arbitrary *half-odd-integer* spin but not to integer spin, thus suggesting a difference between these two cases which was first pointed out by Haldane.[3] In fact

[†]Reprinted from *Phys. Rev. B* (in press) (1988).

this theorem implies[2] that for half-integer-spin, in the infinite length limit, either the ground state is degenerate or else there are gapless excitations. In the former case this ground state degeneracy is likely to be the result of a spontaneously broken symmetry. A unique ground state with a gap is impossible for half-integer-spin. By contrast it *can* occur for integer-spin as was proven rigorously by constructing a solvable model.[4]

Lieb, Schultz and Mattis[1] also pointed out that their theorem could be extended to higher dimension. We wish to give a somewhat more detailed discussion of this extension and draw some (alas non-rigorous) conclusions from it. Besides extending it to higher dimension, we will also discuss its extension to include phonons (Heisenberg-Peierls model). Another extension is to SU(n) generalizations of the usual SU(2) spin systems. In fact the theorem appears to be remarkably insensitive to details of the Hamiltonian, due to its essentially topological nature. In the case of ordinary SU(2) antiferromagnets, it works whenever the total spin per unit cell is half-odd-integer. Thus, for example, it works for half-odd-integer-spin on an arbitrary Bravais lattice (in which the unit cell contains a single spin). These cases include the two-dimensional triangular and square lattices, discussed by Anderson and coworkers[5] in the context of "resonating valence bonds." The theorem fails for a half-odd-integer-spin chain with alternating interaction strength, or for a half-odd-integer-spin honeycomb lattice antiferromagnet, since there are two spins per unit cell in these cases. Indeed a solvable spin-3/2 honey-comb lattice model was studied[4,6] which was proven to have exponentially decaying correlation functions and a unique ground state and appears very likely to have a gap.

In the case of realistic Heisenberg Hamiltonians, a very powerful theorem of Dyson, Lieb and Simon,[7] together with its extension to two dimensions[8] prove that the ground state is Néel ordered for bipartite lattices and sufficiently large spin. For the square, cubic or honeycomb lattice this theorem applies for spin $s \geq 1$ (The triangular lattice is not bipartite).

Haldane developed an approximate mapping of the large-s antiferromagnet onto the σ-model, valid in any dimension.[3] We use this to show the tendency towards Néel order for larger spin, based on renormalization group ideas. This suggests that there should be a phase transition from a Néel ordered to disordered state as a function of the spin-wave coupling constant g which decreases with increasing s and also depends on the strength of next nearest neighbor couplings.[9] In general frustrating spin-couplings tend to increase the spin-wave coupling constant and thus to increase the strength of disordering fluctuations. Naively, the σ-model mapping seems to suggest that the disordered phase should have a unique ground state and a gap, analogous to the high-temperature phase of a classical ferromagnet in one higher dimension. However, the Lieb-Schultz-Mattis theorem[1] implies that this cannot be the case in general. In one dimension this difficulty is removed by the inclusion of a topological term in the σ-model.[9] While a topological term also exists in the two-dimensional case, it is unclear whether it plays a role. Indeed, the σ-model mapping, while generally valid in the weak coupling Néel-ordered phase right up to the critical point, may break down in the strong-coupling phase in some cases. Thus while a transition out of the Néel phase should occur for sufficiently strong spin-wave coupling, it is unclear what the strong coupling phase is. The Lieb-Schultz-Mattis theorem gives valuable information about the possibilities.

The rigorous results on Néel order[7,8] imply that, in the case of Heisenberg antiferromagnets on bipartite lattices, only for $s = 1/2$ is there a chance of a "fluctuation-dominated" ground state. For the square lattice if the ground state is not Néel ordered (finite-size calculations[10] suggest that it is) then the Lieb-Schultz-Mattis theorem suggests that either translational symmetry is broken, leading to a doubling of the unit cell, as in the spin Peierls phase, or else there are gapless excitations. Only the $s = 1/2$ honeycomb lattice seems to be a candidate for an experimentally realizable higher-dimensional extension of the Haldane phase,[3] characterized by a unique ground state and a gap.

The outline of this paper is as follows. In the next section, we will review the Lieb-Schultz-Mattis theorem and discuss its extension to higher dimension and to include phonons. In Section III We will take up the issue of Néel order versus disorder, by considering the large-s mapping onto the non-linear σ-model. In Section IV we will give some further discussion of the implication of these results to triangular-lattice antiferromagnets and to the high-T_c superconductors.

THE LIEB-SCHULTZ-MATTIS (LSM) THEOREM

Consider the $s = 1/2$ Heisenberg antiferromagnet on a periodic chain of length L even.

$$H = \sum_i \mathbf{S}_i \cdot \mathbf{S}_{i+1}$$

We wish to prove that there is a low energy excitation of $O(1/L)$. We may assume that the ground state $|\psi_0\rangle$ is unique, since otherwise the result would be trivially true. In fact, a rigorous proof of uniqueness exists in this case and most others of interest.[11] The proof of a low-energy excitation[1] proceeds by constructing a state $|\psi_1\rangle$ which has low energy, i.e., $\langle\psi_1|(H-E_0)|\psi_1\rangle = O(1/L)$, and which is orthogonal to the ground state. $|\psi_1\rangle$ is constructed by making a unitary transformation on $|\psi_0\rangle$, namely a slowly varying rotation about the z-axis:

$$|\psi_1\rangle = U|\psi_0\rangle,$$
$$U \equiv \exp\left[i(2\pi/L)\sum_n nS_n^z\right]. \tag{1}$$

This state has low energy because, for large L, the relative rotation of two neighboring sites in $O(1/L)$. Requiring the relative rotation between the Lth and 1st sites to be small fixed the overall coefficient in the exponential in U. Noting that

$$\exp(i\theta S^z)S^+ \exp(-i\theta S^z) = \exp(i\theta)S^+,$$

we see that

$$\langle\psi_1|(H - E_0)|\psi_1\rangle = (1/2)\left\{\left[\exp(i2\pi/L) - 1\right]\sum_i\langle S_i^+ S_{i+1}^-\rangle + \text{h.c.}\right\}.$$

Since the ground state is unique it must be isotropic, implying

$$\langle S_i^+ S_{i+1}^-\rangle = \langle S_i^- S_{i+1}^+\rangle.$$

Thus

$$\langle\psi_1|(H - E_0)|\psi_1\rangle = \left[\cos(2\pi/L) - 1\right]\sum_i\langle S_i^+ S_{i+1}^-\rangle.$$

Since $S_i^+ S_{i+1}^-$ is a bounded operator, $\langle\psi_1|(H - E_0)|\psi_1\rangle = textO(1/L)$ in fact we have:

$$\langle\psi_1|(H - E_0)|\psi_1\rangle = \left[\cos(2\pi/L) - 1\right]2E_0/3$$

where E_0 is the ground state energy.

Of course, merely constructing a low energy state proves nothing; it might become equal to the ground state as $L \to \infty$. To complete the proof, we must show that this doesn't happen. This can be done by showing that $|\psi_1\rangle$ is orthogonal to $|\psi_0\rangle$. This is true because $|\psi_1\rangle$ has momentum π relative to the ground state, and thus must be orthogonal due to our uniqueness assumption which implies that $|\psi_0\rangle$ is a momentum eigenstate. To calculate the momentum of $|\psi_1\rangle$, we must calculate the effect on U of a translation by one site:

$$U \to TUT^{-1} = \exp\left[i(2\pi/L)\sum_{n=1}^{L-1} nS_{n+1}^z + LS_1^z\right]$$

$$= U\exp\left[-i(2\pi/L)\sum_{n=1}^{L} S_n^z\right]\exp\left(i2\pi S_1^z\right).$$

The ground state has spin-0 since it is unique, and thus the first exponential gives one:

$$U|\psi_0\rangle \to U\exp(i2\pi S_1^z)|\psi_0\rangle.$$

Finally since S_1^z has eigenvalues $\pm 1/2$,

$$TU|\psi_0\rangle = -UT|\psi_0\rangle.$$

Thus there is a low energy state of momentum π, relative to the ground state. This proof extends immediately to arbitrary half-odd-integer-s^2 but fails miserably in the integer-s case. The reason is that $\exp(i2\pi S_1^z)$ is ± 1 for integer or half-odd-integer s respectively and so the momentum is zero or π respectively. In the former case it cannot be proved that $|\psi_1\rangle$ is orthogonal to $|\psi_0\rangle$. This appears not to be a mere technicality but to cut to the heart of the difference between integer and half-odd-integer spin, using as it does, the fact that half-odd-integer wave functions change sign under 2π rotations.

In the half-odd-integer case, the proof can easily be extended to much more general Hamiltonians. Anisotropic and non-nearest neighbor interactions can be added. The fact that $|\psi_1\rangle$ has low energy is true for essentially *any* reasonable Hamiltonian, and the fact that the momentum is π did not use any property of the Hamiltonian at all, except that translation by one site is a symmetry. (This latter property would fail with alternating interactions.) Of course, the assumption of a unique ground state can fail for some Hamiltonians (for example ferromagnetic ones), in which case, while there *is* a low energy state (another ground state), it doesn't necessarily have momentum π.

Of course even if the finite chain ground state is unique, the infinite length ground state may not be. For the Hamiltonian

$$H = \sum_i \left[\mathbf{S}_i \cdot \mathbf{S}_{i+1} + (1/2)\mathbf{S}_i \cdot \mathbf{S}_{i+2} \right], \tag{2}$$

the finite-chain ground state is two-fold degenerate, the two ground states differing by translation by one site. These two ground states correspond to pairs of nearest neighbor valence bonds.[12,4] It is believed[13] that this two-fold degeneracy persists for a finite range of second nearest neighbor coupling, over some of which the *finite* chain presumably has a unique ground state. In this situation, the ground state of a large finite chain is essentially the symmetric combination of the two infinite chain ground states. The low-energy excited state is the antisymmetric combination. The overlap of the two different pure states and hence the splitting of the finite chain eigenstates is $O[\exp(-\text{const} \cdot L)]$. This low-energy state has momentum π relative to the ground state. One expects a finite gap to all other states, as $L \to \infty$. This has been proven rigorously[4] for the solvable model of Eq. (2).

In the case of the Heisenberg Hamiltonian, the Bethe ansatz solution shows that the ground state is unique but the dispersion relation for one-particle excitations is

$$E \propto |\sin k|.$$

This vanishes at $k = \pi$, as required by the LSM theorem.

The LSM theorem seems to imply that half-odd-integer antiferromagnets are generically in a gapless non-degenerate phase or else have broken translational symmetry.

Let us now consider higher-dimensional generalizations of the LSM theorem. Consider first a half-odd-integer-s Heisenberg model on a square lattice of length L and width M with periodic boundary conditions. Let us again attempt to construct a low-energy state, orthogonal to the ground state.[1] We may again consider making slowly varying rotations of definite momentum. Suppose we use a unitary operator of momentum $(\pi, 0)$:

$$U \equiv \exp \left[i(2\pi/L) \sum_{\mathbf{x}} x S_{\mathbf{x}}^z \right]$$

(the coordinates are integers). We may bound the energy of $U|\psi_0\rangle$ as before. There is no increase in energy for the vertical bonds and each of the LM horizontal bonds has an increase in energy of $O(1/L^2)$. Thus the energy is $O(M/L)$. This is a low energy state for a strip with $L \gg M$. In particular it gives a zero energy state for an infinite strip.

However, we must ask if $U|\psi_0\rangle$ is orthogonal to the ground state. We may calculate the momentum of U as before. We find

$$TUT^{-1} = U \exp\left[-i(2\pi/L)\sum_{\mathbf{x}} S^z_{\mathbf{x}}\right] \cdot \exp\left[i2\pi \sum_y S^z_{(1,y)}\right].$$

As before, we may assume that the ground state has total spin zero. The second exponential contains the total z-component of spin on the first column. This will be an integer or half-odd-integer according to whether M is even or odd, respectively. Thus the theorem works if and only if the number of rows, M, is odd.

This result is certainly much less complete than in the one-dimensional case. The energy is only small if the strip is much longer than it is wide, and furthermore, the number of rows must be odd. We hasten to observe that the result *does* hold with periodic boundary conditions, and that requiring an odd number of rows doesn't cause any obvious pathologies, since the total number of *sites* is even.

There is another respect in which the result is less complete in two dimensions. To actually prove rigorously the existence of ground state degeneracy or zero gap, we constructed[2] a low energy state spread over a region of length ℓ in a chain of length $L \gg 1$, with energy $O(1/\ell)$. This was done by the same procedure, introducing a slow twist over a portion of the chain. In two dimensions, although we *could* twist over only a portion of each row, we find it necessary to twist *all* rows, *i.e.*, the excitation fills the whole width of the lattice. The problem is that, in principle, such a spread out excitation might become completely unobservable for the infinite system without necessarily implying ground state degeneracy. An example of such an excitation would be a spin-wave of momentum $k_y = 0$. This would be no more observable in an infinite lattice than a photon with wave-length equaling the size of the universe.

However, we regard this as a technical limitation not a fundamental one for the following reason. In a physically sensible model, if there are $k_y = 0$ spin-waves, there are also spin-waves with k_y close to zero. By forming a linear combination of these states, we can always obtain a localized low-energy state. On the other hand, if the low-energy $k_y = 0$ state is *not* accompanied by other nearby low-energy states, then it should indicate ground state degeneracy. Likewise, if the state only existed for an odd number of rows, we would again expect broken symmetry. One might be happier if the state could be found for an $L \times L$ lattice. However, we expect that the energy gap should *decrease* not increase, if we increase the width from $M(\ll L)$ to L. While this argument has not been made rigorous, we can think of no physical counter-examples. Of course, this may simply reflect a lack of imagination.

We feel that this theorem, incomplete though it is, *does* imply essentially the same result as in one dimension. Namely, either there are gapless excitations of momentum $(0, \pi)$ in the infinite area limit, or else broken translational symmetry. In any event, any approximate solution of the model must pass the test of having a low-energy state for a long strip with an odd number of rows.

Let us now consider other lattices. The theorem can be extended immediately to a cubic lattice in three dimensions. The energy is $O(A/L)$ where A is the area and L the length. This state has low energy for a very long box. The proof also works for a triangular lattice in two dimensions since an odd number of rows is again consistent with periodic boundary conditions. The proof goes through exactly as before. We may again choose the rotation at location $\mathbf{x}$ to be $(2\pi/L)x$, where the x-axis lies along a lattice row. There is now a contribution to the energy of $O(1/L^2)$ from all three types of bonds, again giving a total energy of $O(M/L)$. It was crucial that we could choose an odd number of rows consistently with periodic boundary conditions. Note that the fact that the triangular lattice is not bipartite plays no role here. In fact, it is more natural to take an odd number or rows in this case. For a square lattice, an odd number of rows would force a seam into the Néel state. (Of course this is not a problem for our arguments since the Néel phase has broken translational symmetry.) For the triangular lattice, an odd number of rows is consistent with a tripartite ordered state, provided that that number is divisible by three. Now consider the honeycomb lattice with half-odd-integer spin. We may again consider making a rotation whose magnitude varies with distance along a lattice row. But in this case, it is not possible to have an odd number of rows consistent with periodic boundary conditions. This follows because there are two inequivalent types of rows and neighboring rows are not connected by primitive translations. Indeed an $s = 3/2$ model was constructed which can be proven to have exponential decaying correlation functions[4] and a unique ground state[6] and appears to have a gap.

Let us now consider the extension of the theorem beyond Hamiltonians containing only spin variables. We may easily generalize it to a Heisenberg-Peierls model in which the magnetic interaction strength between two neighboring spins, J_{ij} depends in some way on the distance between them, the displacement of each spin from its equilibrium position being treated as a quantum variable. The proof proceeds exactly as before. The unitary transformation has no effect on the phonon degrees of freedom. The excitation energy for a chain is given by

$$\langle \psi_1 | (H - E_0) | \psi_1 \rangle = \left[\cos(2\pi/L) - 1 \right] \sum_i \langle J_{i,i+1} S_i^+ S_{i+1}^- \rangle$$

and so should $O(1/L)$. The theorem was also extended to chains with $SU(n)$ generalizations of $SU(2)$ spin variables on the sites.[2] The extension of these cases to higher dimension goes exactly the some way as for $SU(2)$. A large-n limit of the square lattice $SU(n)$ Heisenberg model was solved recently,[14] and that solution was consistent with our conclusion, namely translational symmetry was broken in the ground state.

We have *not* been able to extend the proof to itinerant electron models although we expect that the result holds in these cases as well. Field theory analysis of one-dimensional models suggests that in the phase in which both the spin and charge excitations have a gap, there is a broken translational symmetry. The large-n solution[14] of the Hubbard-Heisenberg model on a square lattice is also consistent with the general conclusion.

σ -MODEL MAPPING

The one-dimensional Heisenberg model does not display Néel order even at $T = 0$, due to infrared-singular quantum fluctuations. (This is the quantum version of the Mermin-Wagner theorem,[15] known in field theory as Coleman's theorem.[16]) On the other hand, in two dimensions Néel order is possible in the ground state, although not inevitable. A nice way of understanding these issues is to make a mapping of the spin system onto the non-linear σ-model.[3,17,9]

In the one-dimensional case, this can be done[17,9] by combining pairs of neighboring spins to define the field and rotation generators of the σ-model:

$$\phi(2n + 1/2) \equiv (\mathbf{S}_{2n} - \mathbf{S}_{2n+1})/2\sqrt{s(s + 1)}$$
$$\mathbf{l}(2n + 1/2) \equiv (\mathbf{S}_{2n} + \mathbf{S}_{2n+1})/2.$$

ϕ and $\mathbf{l}$ are then assumed to be slowly varying on the scale of the lattice spacing; we are keeping only momentum modes near zero and π, the two low-energy regions for an antiferromagnet. In the large-s continuum limit, φ and $\mathbf{l}$ obey the commutation relations and constraints of the σ-model:

$$\left[l^a(x), \phi^b(y)\right] = i\varepsilon^{abc}\phi^c(x)\delta(x - y)$$
$$\left[l^a(x), l^b(y)\right] = i\varepsilon^{abc}l^c(x)\delta(x - y)$$
$$\left[\phi^a(x), \phi^b(y)\right] = i\varepsilon^{abc}l^c(x)\delta(x - y)/s(s + 1) \to 0$$
$$\phi \cdot \mathbf{l} = 0$$
$$\phi^2 = 1 - \mathbf{l}^2/s(s + 1) \to 1.$$

Ignoring higher derivative terms, the Hamiltonian density becomes

$$H/v = (g/2)[\mathbf{l} + (\theta/4\pi)(d\phi/dx)]^2 + (1/2g)(d\phi/dx)^2,$$

with velocity, coupling constant and topological angle

$$v = 2\sqrt{s(s + 1)}, \quad g = 2/\sqrt{s(s + 1)}, \quad \theta = 2\pi\sqrt{s(s + 1)}.$$

The corresponding Lagrangian density is:

$$L = (1/2g)(\partial_\mu \phi)^2 + (\theta/4\pi)\phi \cdot (\partial_\mu \phi \times \partial_\nu \phi)\varepsilon^{\mu\nu}.$$

θ multiplies the topological term which is always i times an integer, the winding number, for a smooth configuration in Euclidean space. Thus θ is a periodic variable, and the different behavior for s integer or half-integer can be explained. For $\theta = 0$, s integer, we simply have the Lagrangian describing the continuum limit of a *two-dimensional* classical ferromagnet, with temperature $g = 2/\sqrt{s(s+1)}$. For large s the coupling constant is weak and we may do perturbation theory. This is done by assuming that ϕ has an expectation value, corresponding to Néel order. The perturbative spectrum consists of two Goldstone bosons. The leading infrared divergences of σ-model perturbation theory would correspond to standard spin-wave perturbation theory based on the Holstein-Primakoff approximation. However, we find that the coupling grows with increasing length scale (or decreasing energy scale):

$$dg/d\ln L = g^2/2\pi.$$

This suggests that the symmetry is not really spontaneously broken, and that there is a finite correlation length of order

$$\xi \approx \exp\left[\pi\sqrt{s(s+1)}\right]$$

and a corresponding gap $\Delta \approx v/\xi \approx s \cdot \exp\left[-\pi\sqrt{s(s+1)}\right]$. This behavior corresponds to the fact that the critical temperature of the two-dimensional classical ferromagnet is zero, and there is an exponentially large correlation length at low temperatures.

The behavior of the $\theta = \pi$ σ-model is less familiar, but there appears to be unbroken symmetry with a (non-Goldstone) massless sector.

All of this holds for arbitrarily large s. We may also add a second nearest neighbor coupling. This modifies the σ-model coupling constant to[9]

$$g = 2/\left[\sqrt{s(s+1)}\sqrt{1-4J_2}\right].$$

A frustrating coupling ($J_2 > 0$) tends to increase the σ-model coupling, disfavoring the Néel-ordered state.

We may attempt to repeat the above procedure in higher dimension. The validity of the σ-model mapping in higher dimension was pointed out by Haldane.[3] Consider first the simplest case of a square lattice in two dimensions. The natural extension of the above approach, is to define continuum limit variables on every fourth plaquette. We may define a σ-model field, the order parameter:

$$\phi\left(2x+\frac{1}{2}, 2y+\frac{1}{2}\right) \equiv (S_{2x,2y} - S_{2x+1,2y} - S_{2x,2y+1} + S_{2x+1,2y+1})/4\sqrt{s(s+1)}.$$

We may also define the continuum rotation generator, related to the conjugate momentum for ϕ, as:

$$1\left(2x+\frac{1}{2}, 2y+\frac{1}{2}\right) \equiv (S_{2x,2y} + S_{2x+1,2y} + S_{2x,2y+1} + S_{2x+1,2y+1})/4.$$

ϕ and $\mathbf{l}$ again obey the correct commutation relations and constraints for large s. However a difference emerges from the one-dimensional case. To conserve the number of degrees of freedom, we must define two other fields:

$$\mathbf{A}_x\left(2x+\frac{1}{2},2y+\frac{1}{2}\right)\equiv(\mathbf{S}_{2x,2y}-\mathbf{S}_{2x+1,2y}+\mathbf{S}_{2x,2y+1}-\mathbf{S}_{2x+1,2y+1})/4[s(s+1)]^{1/4}$$

$$\mathbf{A}_y\left(2x+\frac{1}{2},2y+\frac{1}{2}\right)\equiv(\mathbf{S}_{2x,2y}+\mathbf{S}_{2x+1,2y}-\mathbf{S}_{2x,2y+1}-\mathbf{S}_{2x+1,2y+1})/4[s(s+1)]^{1/4}$$

Treating all these fields as slowly varying, $\phi, \mathbf{l}, \mathbf{A}_x$ and $\mathbf{A}_y$ correspond to the Fourier modes of the spin operators, $\mathbf{S}$, with momentum near (π,π), $(0,0)$, $(\pi,0)$ and $(0,\pi)$ respectively. $\mathbf{l}$ generates rotations of the $\mathbf{A}_i$ as well as of ϕ. In the continuum limit ϕ commutes with the $\mathbf{A}_i$:

$$[\phi^a(\mathbf{x}),A_x^b(\mathbf{y})]=i\varepsilon^{abc}A_y^c(\mathbf{x})\delta(\mathbf{x}-\mathbf{y})/\sqrt{s(s+1)}\to 0$$
$$[\phi^a(\mathbf{x}),A_y^b(\mathbf{y})]=i\varepsilon^{abc}A_x^c(\mathbf{x})\delta(\mathbf{x}-\mathbf{y})/\sqrt{s(s+1)}\to 0.$$

On the other hand, the $\mathbf{A}_i$ have a non-zero commutator:

$$[A_y^a(\mathbf{x}),A_x^b(\mathbf{y})]=i\varepsilon^{abc}\phi^c(\mathbf{x})\delta(\mathbf{x}-\mathbf{y}),$$
$$[A_x^a(\mathbf{x}),A_x^b(\mathbf{y})]=[A_y^a(\mathbf{x}),A_y^b(\mathbf{y})]i\varepsilon^{abc}l^c(\mathbf{x})\delta(\mathbf{x}-\mathbf{y})/\sqrt{s(s+1)}\to 0.$$

All four fields are exactly mutually orthogonal. They also obey the constraint:

$$\phi^2=1-\mathbf{l}^2/s(s+1)-\mathbf{A}_x^2/\sqrt{s(s+1)}-\mathbf{A}_y^2/\sqrt{s(s+1)}\to 1.$$

The two additional fields that we have been forced to introduce do not seem to have any obvious interpretation in the σ-model. Making a gradient expansion, we now find the Hamiltonian density:

$$H/v=(g/2)\mathbf{l}^2+(1/2g)(\nabla\phi)^2+(1/2)\mathbf{A}_x^2+(1/2)\mathbf{A}_y^2,$$

where $g=2/\sqrt{s(s+1)}$, $v=4\sqrt{s(s+1)}$. Once again, g will increase with a frustrating second-nearest-neighbor coupling. Temporarily ignoring the extra fields, let us consider the σ-model alone. We now have a continuum version of the *three-dimensional* classical ferromagnet at temperature g. Thus there should be a phase transition at some finite value of g (or order one). This can be seen, for example from the $(2+\varepsilon)$ expansion of the σ-model.[18] The fixed point at couplings of order ε presumably persists up to three dimensions. In the weak coupling phase, there is Néel order and two Goldstone bosons.

Let us now consider the affect of the additional fields, in the weak coupling phase. Choosing $\langle\phi^a\rangle=\delta^{a3}$, the commutation relations become, to leading order in $1/s$,

$$[A_x^1(\mathbf{x}),A_y^2(\mathbf{y})]=[A_x^2(\mathbf{x}),A_y^1(\mathbf{y})]=-i\delta(\mathbf{x}-\mathbf{y}),$$

with the other commutators lower order. Thus (A_x^1, A_y^2) and $(A_x^2, -A_y^1)$ define two field-conjugate momentum pairs, while A_x^3 and A_y^3 are classical fields. In this approximation, the extra fields are decoupled and massive. Thus the low-energy sector in the Néel phase consists only of the Goldstone bosons as expected. Once again, the Goldstone modes, with momentum near (π, π), and $(0, 0)$, are those obtained from standard spin-wave theory. The extra fields $\mathbf{A}_i$, with momentum near $(\pi, 0)$ and $(0, \pi)$, don't affect the leading infra-red behavior of perturbation theory, or, presumably the existence of a critical point. However, it is much less obvious what role these extra fields may play in the strong-coupling phase.

Based on the LSM theorem and our experience with the one-dimensional case, it seems likely that the nature of the strong-coupling phase depends radically on whether the spin is integer or half-odd-integer. In the former case a unique ground state with a gap may occur, corresponding to the disordered phase of a classical three-dimensional ferromagnet. However, in the half-odd-integer case, this is inconsistent with the LSM theorem. Instead there is presumably either a breaking of the symmetry of translation by one site, or else a unique ground state with vanishing gap. The former case corresponds to some two-dimensional generalization of the dimerized phase. The lattice translational symmetry is broken but not the spin-rotational symmetry. In the latter case we can imagine (at least) two possibilities. One is that there is a gapless pseudo Fermi surface and effectively massless free fermions as in the resonating valence bond model discussed by Anderson and collaborators.[5] Another possibility, is that the gap vanishes only at discrete points in momentum space, probably $(0, 0)$, $(0, \pi)$, $(\pi, 0)$ and (π, π). In this case, here is likely to be a $(2+1)$-dimensional field theory description of the gapless sector. This *cannot* be described purely by the σ-model, since in his case the gap would only vanish at $(0, 0)$ and (π, π), contradicting the LSM theorem. The fields $\mathbf{A}_i$ must also play a role in the continuum limit. It is possible that the Hopf topological term of the $(2+1)$-dimensional σ-model[19] appears. On the other hand, if there is a gapless Fermi surface in the strong-coupling phase, then a $(2+1)$-dimensional field theory is *not* the correct description, since for such a field theory the gap would only vanish at discrete points in momentum space.

In the large-n limit of the Heisenberg model,[14] we found a dimerized ground state, although a non-degenerate ground state with the gap vanishing at discrete points in momentum space, represented another locally stable state of slightly higher energy. (For the Heisenberg-Hubbard model, the latter state became the ground state for a range of parameters and a ground state with a gapless Fermi surface also occurred away from half-filling.)

In renormalization group language, there is an infrared unstable fixed point corresponding to the Néel ordering transition, as a function of g (which can be controlled by varying s or a second-nearest neighbor coupling). For integer-s the flow on the strong-coupling side is to some short-range (non-universal) fixed point.

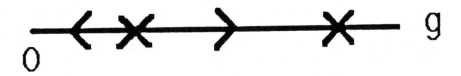

FIGURE 1 The proposed renormalization group flow diagram for $d \geq 2$ antiferromagnets subject to the LSM theorem.

For half-odd-integer-s the flow is to some other attractive fixed point (see Fig. 1), or else to some short-range (non-universal) dimerized phase.

The above discussion was given for a square lattice, but very similar results would emerge for *any* bipartite lattice, in any dimension greater than one. In the continuum limit, the σ-model fields ϕ and $\mathbf{l}$ would arise together with a lattice-dependent number of additional fields, which would be massive in the Néel phase. A transition out of the Néel phase should be generic for sufficiently small s or large second-nearest neighbor coupling. This is the analogue of the fact that a classical ferromagnet has a finite temperature transition for any dimension greater than two. In cases where the LSM theorem applies, the strong-coupling phase should be either dimerized or gapless; in other cases it may be a short-range non-degenerate phase with a gap.

There does not appear to be any simple way of estimating the critical value of s at which the Néel transition occurs, for various lattice types. The critical coupling in the σ-model is non-universal and regularization dependent. The most useful result, in this regard, is the rigorous theorem of Dyson, Lieb and Simon,[7] which shows that there is Néel order for the Heisenberg Hamiltonian and any $s \geq 1$, for most lattice types.

IV. DISCUSSION

Let us summarize our main conclusions. In any dimension greater than one and any lattice type, there should be a critical coupling separating the Néel phase from a strong-coupling phase, where the σ-model coupling constant decreases as $1/s$ but increases with a frustrating second-nearest-neighbor spin coupling. In cases where the LSM theorem does not apply, namely where the spin per unit cell is integer, a unique ground state with a gap may occur. Where the LSM theorem *does* apply we expect the strong-coupling phase to have either broken translational symmetry or else gapless excitations. In the latter case, these may either be on a Fermi surface, or at the discrete points $(0,0)$, $(0,\pi)$, $(\pi,0)$ and (π,π) only.

What are the prospects for testing these predictions? Numerical work on finite two-dimensional lattices is certainly one possibility. However we are not optimistic that large enough systems are manageable to really resolve the issues. In the one-dimensional case it was necessary to go[19] to spin-one chains of length 30 to show fairly convincingly the existence of a gap, although the experts now generally agree that a sufficiently sophisticated analysis of a chain of 16 sites might really have been enough. Analysis of two-dimensional antiferromagnets, even for the $s = 1/2$ case, will presumably be restricted to sizes of about 6 × 6. Given the history of the one-dimensional controversies, the situation does not seem very favorable for clear answers.

The solvable $s = 3/2$ honeycomb lattice model of [4], and [6] provides an example of the strong-coupling phase with a unique ground state and (presumably) a gap.

Most interesting is the possibility of experimentally observing the strong-coupling phase in a quasi-two-dimensional (or three-dimensional) antiferromagnet. Nature restricts us, more or less, to pure nearest neighbor Heisenberg Hamiltonians. The powerful results of Dyson, Lieb and Simon[7] show that the critical value of s is less than one. It may be less than 1/2 also, meaning that even the $s = 1/2$ case is Néel ordered. This could depend on lattice type, of course. The $s = 1/2$ square lattice is subject to the LSM theorem and so should have a broken translational symmetry or zero gap in the strong-coupling phase. We might expect quantum fluctuations to be a even stronger for the $s = 1/2$ honeycomb lattice, since the number of nearest neighbors is only three. For this case the strong-coupling phase may have a unique ground state with a gap.

Let us summarize the situation for the triangular lattice antiferromagnets. A tripartite magnetically ordered state is a possibility. The Dyson, Lieb, Simon theorem[7] has not been extended to tripartite lattices, so no rigorous results are known on this ordered state. The σ-model mapping could presumably be carried out, and one would expect order for sufficiently large spin, and disordered phases for small enough spin and large enough frustrating second-nearest-neighbor couplings. Since the LSM theorem applies for half-odd-integer-s, the strong-coupling phase should have either broken translational symmetry or vanishing gap in this case. An interesting variational ground state wave-function for the $s = 1/2$ triangular case was recently discussed by Kalmeyer and Laughlin,[20] based on Laughlin's fractional quantum Hall effect wave-function.[21] They used a boson representation for the spin variables with an infinite hard-core repulsion. An empty or occupied site corresponds to $S_z = -1/2$ or $1/2$ respectively. In the boson representation, U becomes:

$$U = \exp\left[i(2\pi/L)\sum_j x_j\right]$$

The energy is $O(M/L)$ as can be seen by applying U to the boson hopping term, representing the S^+S^- couplings. Under a translation by one site, each x_j is increased by 1 (or decreased by $L-1$ if $x = L$). The total number of bosons is $LM/2$ so $U \to -U$ for M odd. Thus this excited state has momentum $(\pi, 0)$ relative to the ground state.

The CuO_2 planes in La_2CuO_4 (and also in $YBa_2Cu_3O_x$ for some values of x) are probably well described by the Heisenberg model. An orthorhombic distortion of the lattice has been observed.[23] However, apparently the square Cu plaquettes of the tetragonal phase are simply tilted into a rhombus with no breaking of the translational symmetry in the effective two-dimensional Heisenberg model. Néel order has also been observed.[24,25] However it is not clear if the two-dimensional planes would order at $T = 0$ or if the observed ordering is entirely due to interplane coupling. Some experimental observations relevant to this question are presented in [26]. In the latter case the material may be well described by a disordered $s = 1/2$ square lattice ground state. The LSM theorem should then imply either broken translational symmetry or vanishing gap. Apparently, no indications of broken translational symmetry *in the effective Heisenberg model* have been observed. If the gap vanishes linearly at discrete points in momentum space, as in a relativistic $(2+1)$-dimensional quantum field theory, then the specific heat would be quadratic at low T. On the other hand, if the gap vanishes linearly on a Fermi surface then the specific heat would be linear at low T, fitting experimental observations.[27]

ACKNOWLEDGEMENTS

I would like to thank P. W. Anderson, C. Kallin, T. Kennedy, S. Kivelson, E. Lieb, and H. Tasaki for very useful discussions.

REFERENCES

1. E. Lieb, T. Schultz, and D. Mattis, *Ann. Phys. (N.Y.)*, **16**, 407 (1961).
2. I. Affleck and E. Lieb, *Lett. Math. Phys.*, **12**, 57 (1986).
3. F. D. M. Haldane, *Phys. Lett.* , A93, 464 (1983), *Phys. Rev. Lett.*, **50**, 1153 (1983).
4. I. Affleck, T. Kennedy, E. Lieb, and H. Tasaki, *Phys. Rev. Lett.*, **59**, 799 (1987), and (1987) Princeton preprint.
5. P. W. Anderson, *Res. Bull.*, **8**, 153 (1973); P. Fazekas and P. W. Anderson, *Philos. Mag.*, **30**, 432 (1974); P. W. Anderson, *Science*, **235**, 1196 (1987); G. Baskaran, Z. Zou, and P. W. Anderson, *Solid state Commun.*, **63**, 973 (1987) and this volume; P. W. Anderson, G. Baskaran, and Z. Zou, to be published;

P. W. Anderson, G. Baskaran, Z. Zou, and T. Hsu, *Phys. Rev. Lett.*, **58**, 2790 (1987); Z. Zou, and P. W. Anderson, *Phys. Rev. B*, in press and this volume; G. Baskaran and P. W. Anderson, Princeton preprint.

6. T. Kennedy, E. Lieb, and H. Tasaki, private communication.
7. F. J. Dyson, E. H. Lieb, B. Simon, *J. Stat. Phys.*, **18**, 335 (1978).
8. E. Jordao Neves and J. Fernando Peres, *Phys. Lett.*, **A114**, 331 (1986).
9. F. D. M. Haldane, unpublished; I. Affleck, *Nucl. Phys.*, **B257**, 397 (1985).
10. J. Oitmaa and D. D. Betts, *Can. J. Phys.*, **56**, 897 (1978).
11. E. Lieb and D. Mattis, *J. Math. Phys.*, **3**, 749 (1962).
12. C. K. Majumdar and D. K. Ghosh, *J. Math. Phys.*, **C3**, 911 (1970).
13. F. D. M. Haldane, *Phys. Rev. B*, **25**, 4925 (1982).
14. I. Affleck and J. B. Marston, (1987) U.B.C. preprint.
15. N. Mermin and H. Wagner, *Phys. Rev. Lett.*, **17**, 133 (1966).
16. S. Coleman, *Comm. Math. Phys.*, **31**, 259 (1973).
17. L. Fadeev and L. Takhtajan, unpublished.
18. E. Brezin and J. Zinn-Justin, *Phys. Rev. Lett.*, **36**, 691 (1976); *Phys. Rev. B*, **14**, 3110 (1976).
19. F. Wilzek and A. Zee, *Phys. Rev. Lett.*, **51**, 2250 (1983).
20. H. J. Blote, M. P. Nightingale, *Phys. Rev. B*, **33**, 659 (1986).
21. V. Kalmeyer and R. B. Laughlin, unpublished.
22. R. B. Lauphlin, *Phys. Rev. Lett.*, **50**, 1395 (1983).
23. J. D. Jorgensen, H.-B. Schuttler, D. G. Hinks, D. W. Capone II, K. Zhang, M. B. Brodsky, and D. J. Scalapino, *Phys. Rev. Lett.*, **58**, 1024 (1987).
24. D. Vaknin, S. K. Sinha, D. E. Moncton, D. C. Johnston, J. M. Newsam, C. R. Safinya, and H. E. King, Jr., *Phys. Rev. Lett.*, **58**, 2802 (1987).
25. N. Nishida, H. Miyatake, D. Shimada, S. Ohkuma, M. Ishikawa, T. Takabatake, Y. Nakazawa, Y. Kuno, R. Keitel, J. Brewer, T. Riseman, D. Williams, Y. Watanabe, T. Yamazaki, K. Nishiyama, K. Nagamine, E. Ansaldo, S. Dodds, D. Harshman, and E. Torikai (1987) Triumf preprint.
26. G. Shirane, Y. Endoh, R. J. Birgeneau, M. A. Kastner, Y. Hidaka, M. Oda, M. Suzuki, and T. Murakami (1987) preprint.
27. N. E. Phillips, R. A. Fisher, S. E. Lacy, C. Marcenat, J. A. Olsen, W. K. Ham, and A. M. Stacy, (1987) L. B. L. preprint; E. Airniebl, J. Willis, J. Thompson, C. Huang, and J. Smith (1987) Los Alamos preprint.

HUBBARD-LIKE MODELS

EXPLICITLY INCLUDING OXYGEN

C. M. Varma, S. Schmitt-Rink
AT&T Bell Laboratories
Murry Hill, New Jersey 07974

and

Elihu Abrahams
Serin Physics Laboratories
Rutgers University
Piscataway, New Jersey 08855

Charge Transfer Resonances and Superconductive Pairing in the New Oxide Metals†

INTRODUCTION

All theoretical discussions of the superconductivity in the new oxide metals[1-6] have been within the BCS framework of two-particle pairing. The issue in question is what is the pairing mechanism? Do the Cooper pairs have the usual BCS s-wave symmetry? Is the pairing so strong that the Blatt-Schafroth type ideas of Bose condensation of tightly bound pairs are realized? The points of view proposed may be classified as follows:

I. Pairing through electron-phonon interactions
   (i) conventional[7]
   (ii) bipolaron formation and Bose condensation[8]
II. Pairing through magnetic correlation of the electrons
   (i) exchange of antiferromagnetic (AFM) spin fluctuations[9,10]
   (ii) resonating valence bond (RVB) or spin-bipolaron formation and Bose condensation[11,12]
III. Pairing through exchange of electronic polarization resonances[13]

†Reprinted from *Proceedings of the International Conference on Superconductivity*, ed. S. E. Wolf and V. Z. Kresin, Plenum, New York, 1987.

ELECTRON-PHONON MODELS

Estimates were given earlier of the upper limit on superconducting transition temperatures expected of the new compounds based on their electronic structure.[13] An optimistic upper limit is 30 K. The recent experiments yielding only a very small isotope shift imply that the contribution of electron-phonon interactions to T_c is in fact much smaller.[14,15] Although there are calculations with electron-phonon interactions which give T_c in $La_{2-x}Sr_xCuO_4$ in agreement with experiments, they also predict structural transformations which are not observed.[7] This means there is something crucial missing in the electron-phonon plus band structure models for these materials.

In the bipolaron models, pairs of electrons interact so strongly via their induced lattice deformation that they form real space pairs.[16] This may be modelled by a local effective electron-electron attraction U, $|U| \gg t$, where t is the electron hopping amplitude. The kinetic energy of a pair (which in the strong binding limit is a boson) is $t^2/|U|$ corresponding to an effective mass $|U|/t^2$. Bose condensation occurs at a temperature characteristic of such an effective mass, i.e., $T_c \sim t^2/|U|$. In Fig. 1 we have sketched T_c/t as a function of $t/|U|$.[17] Actually, the renormalization of the hopping amplitude puts an ever stronger limit on T_c due to bipolaron condensation. The weak coupling portion of the curve ($t \gg |U|$) corresponds to the BCS result $T_c \sim t \exp(-t/|U|)$. In the bipolaronic region, the Bose line for T_c/t lies below the BCS extrapolation which, as remarked above, has a T_c of at most 30 K for optimistic parameters of the materials in question.

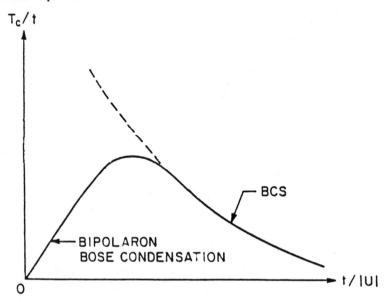

FIGURE 1 Critical temperature T_c as a function of electron hopping amplitude t and on-site attraction U for an attractive Hubbard model.

MAGNETIC MODELS FOR SUPERCONDUCTIVITY

The starting inspiration of such models is that the stoichiometric compound La_2CuO_4, which in band theory should be metallic with a half-filled band, is in fact an insulator and, below about 220 K, an antiferromagnet.[18,19] It is then quite natural to regard this compound as an antiferromagnetic Mott-Hubbard insulator and seek magnetic mechanisms for the superconductivity away from half-filling.

It is absolutely vital that any model for superconductivity accounts as well for the properties at half-filling. We shall see that this does not mean one is irrevocably lead to the magnetic models of superconductivity. The magnetic models all assume that these materials are describable by the Hubbard model. We shall soon discuss that this is not reasonable.

A theory for superconductivity in heavy fermions through exchange of AFM spin fluctuations is quite successful in explaining their properties.[9,20] It predicts an anisotropic gap and a density of states of quasiparticles which unlike BCS is linear in temperature. The quasiparticle properties and the gap in the new superconductors appear to us to be quite conventional. This would appear to rule out such a magnetic mechanism. Experimentally, the evidence for such a BCS-like gap is emerging. The temperature dependence of the London penetration depth is also consistent with BCS-type pairing.[21] The unusual linear specific heat in the superconducting state as well as the linear resistivity in the normal state can be understood as arising from tunneling states for which there is now independent evidence from very low temperature acoustic measurement.[22,23]

MODEL FOR THE OXIDES

Let us first consider the simple transition metal oxides. The Mott insulating phase, the Hubbard model and the super-exchange process leading to antiferromagnetism were developed to understand their properties. A tight-binding model for these is:

$$H = \sum_i (\epsilon_A n_{Ai} + \epsilon_B n_{Bi} + U_A n_{Ai\uparrow} n_{Ai\downarrow} + U_B n_{Bi\uparrow} n_{Bi\downarrow})$$
$$+ \sum_{\langle i,j \rangle} [(V n_{Ai} n_{Bj} + t \sum_\sigma (C_{Ai\sigma}^+ C_{Bj\sigma} + \text{h.c.})] \tag{1}$$

where A stands for the metal ion say and B for the oxygen. V is derived from the Coulomb interaction; it leads to the stability of ionic configurations.

In many cases a truncated version of (1), the Hubbard model, is adequate. The considerations leading to it are as follows: consider the mean field solution of (1) or equivalently some self-consistent one-electron theory, such as Hartree-Fock or density functional theory, etc. This will include the ionic and the covalent

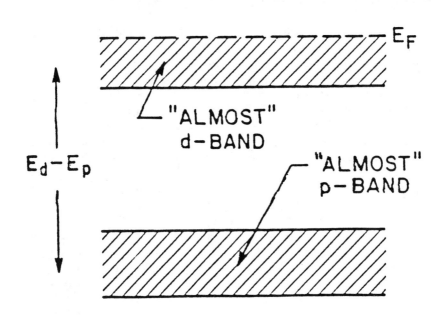

FIGURE 2 Electronic structure of transition metal oxide compounds.

interactions among the ions and give rise to a self-consistent band structure which usually has the form of a set of filled bands which are mostly oxygen p with a small amount of covalent d admixture and unfilled bands which are mostly transition metal d with a small amount of oxygen p admixture (see Fig. 2). The formal O^{--} state has of course been stabilized by the electrostatic interaction between the atoms.

When we now begin to consider corrections beyond one-electron (mean field) theory, it is enough to consider an effective Hamiltonian describing the upper band only provided no term in the Hamiltonian causes significant resonant transitions between the lower and the upper bands. Thus a kinetic energy term in the space of the "d" orbitals and a local repulsion U in this subspace suffices:

$$H = \sum_i \left[\bar{t} \sum_\sigma (C_{i\sigma}^+ C_{i+1\sigma} + \text{h.c.}) + \overline{U} n_{i\uparrow} n_{i\downarrow} \right] \qquad (2)$$

where $\bar{t} \approx \tilde{t}^2/(\tilde{\epsilon}_A - \tilde{\epsilon}_B)$. If $\tilde{t}/(\tilde{\epsilon}_A - \tilde{\epsilon}_B)$, $U/(\tilde{\epsilon}_A - \tilde{\epsilon}_B)$ and $V/(\tilde{\epsilon}_A - \tilde{\epsilon}_B)$ are much smaller than 1, this picture is adequate. If they are $\gtrsim 1$, it is not. Here, $\tilde{\epsilon}_A, \tilde{\epsilon}_B$ and $\tilde{t}$ are the mean field renormalized levels and Cu-O hopping amplitude, respectively.

We shall see that for the oxide superconductors it is not adequate. This can be shown starting either from a band point of view or a localized ionic point of view. Self-consistent band structure calculations have been done for both $La_{2-x}Sr_xCuO_4$

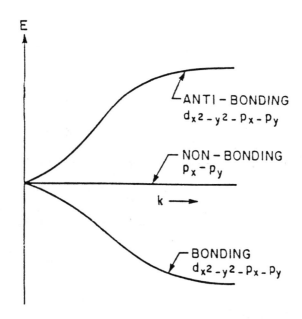

FIGURE 3 Self-consistent one-electron energies for planar CuO_2 networks.

and $YBa_2Cu_3O_7$. A tight-binding fit to the band structure in the plane (as well as along the chains in $YBa_2Cu_3O_7$) in terms of effective atomic levels E_d, E_p and a transfer integral t yields $E_d \approx E_p$![24,25] The band structure has the schematic form shown in Fig. 3. The covalency of the d-p orbitals at the self-consistent ionic state is 100%, there is no gap at all and a dispersionless non-bonding $p_x - p_y$ band lies right at the foot of the conduction band. Any fluctuations involving U or the Cu-O coulomb interaction V dynamically mix the p and d states. Restricting attention to a Cu-like conduction band alone to deduce a Hubbard model is quite useless.

This point of view is reinforced by looking at the problem from a localizeed point of view following Zaanen, Sawatzky and Allen.[26] Consider the two parameters U_A and E_x, where E_x is the charge transfer excitation energy corresponding to $A^n B^m \rightarrow A^{n+1} B^{m-1}$. The relevant E_x for NiO for example is the energy difference $E(Ni^+iO^-) - E(Ni^{++}O^{--})$. In the band picture, Fig. 3, E_x corresponds approximately to the average energy needed to create a hole in the lower bands and an electron in the conduction band taking into account the effective attraction between the electron and the hole, $i.e.$, the excitonic effect.

For $E_x \gg U_A$, see Fig. 4, the Hubbard model is adequate. Within this sector, for $U_A \gg W$, the Mott-insulator is the ground state (W is the conduction band width); the lowest particle-hole excitation corresponds to $2d^n \rightarrow d^{n+1} + d^{n-1}$. But for $E_x \ll U_A$, completely different behavior occurs: for $E_x \gg W$ in this regime, the material is again an insulator, but with the lowest excitation of the charge transfer variety $M^{++}O^{--} \rightarrow M^+O^-$. For $E_x \ll W$ the material is a metal even for very large U_A since fluctuations $M^{++}O^{--} \rightarrow M^+O_-$ are mixed quantum-mechanically. For $E_x \gg W$, if M^{++} has local moments, the material will doubtless order magnetically at a

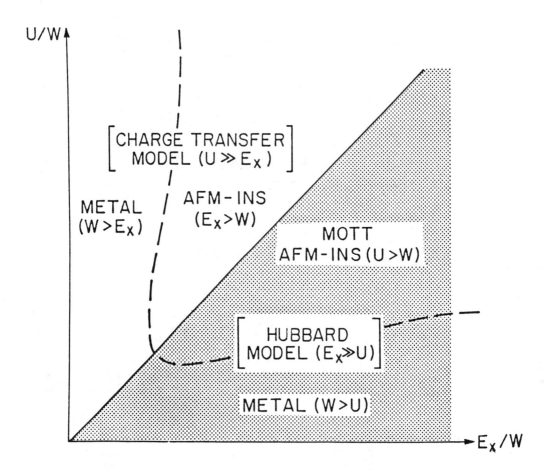

FIGURE 4 Ground state phase diagram of transition metal oxide compounds.

temperature of order $(t^2/E_x)^2/U_A$. (In a simple cubic lattice, because of nesting, this will always be the case, no matter how large W.)

As we move to the right of the periodic table, the ionization potential of the transition metals drops making E_x smaller. Even for NiO, the classic Mott-insulator, Sawatzky and Allen have presented strong evidence that a description purely in terms of a Hubbard model is inadequate.[27] The charge transfer excitations are then even more important in CuO. Some experimental evidence for this is available. This means that the conductivity gap is dominated by the process $Cu^{++}O^{--} \rightarrow Cu^{+}O^{-}$, whereas in a conventional Mott insulator O^{--} would change only virtually and the conductivity process would be $2Cu^{++} \rightarrow Cu^{+++} + Cu^{+}$. Imagine now that we have a material in which, as in the new oxide superconductors, some of the Cu is in the Cu^{3+} state. The excitonic energy for the Cu^{3+} ion is considerably lower than for Cu^{2+} since the electron affinity energy of Cu^{3+} is much larger than that of Cu^{2+}. There is then the possibility of the material being metallic.

In fact, there is also the real possibility of doping producing O_2^{2-} rather than Cu^{3+}; the peroxide ion is known to occur for *excess* oxygen in La_2NiO_4. But La_2CuO_4 forms with oxygen deficiency—in this situation the formation of the peroxide ion on doping is unlikely. The Pauling-Zachariasen rules, when applied to the measured lattice constants, give strong evidence for Cu^{3+}, as does chemical analysis. The charge transfer energy $Cu^{3+}O^{--} \to Cu^{2+}O^{-}$ is however bound to be much smaller than U making the Hubbard model quite a wrong starting point for the discussion of these materials. If the energy E_x is very small compared to U, U may be dropped altogether out of the problem (away from half-filling) simultaneously discarding any matrix elements for $2d^n \to d^{n+1} + d^{n-1}$ in the kinetic energy. Metallic conduction now occurs through the process

$$+ + +, - -, + +$$
$$\to + +, \quad -, \quad + +$$
$$\to + +, \quad - -, + + +$$

etc., in which only intermediate states costing energy E_x are used. E_x here is characteristic of $(+ + +, --) \to (++, -)$ and not $(++, --) \to (+, -)$ which would be much higher in energy.

It is worth noting that neither the band description emphasizing strong covalency nor the ionic description are likely to be completely correct. They are useful only as starting points for discussion . In particular, different experiments— EXAFS, photoemission, lattice constants etc.—are likely to give different values for "valence" or the ionic state of Cu and oxygen if interpreted too literally. We talk here of "Cu^{+++}" merely to illustrate a point; our arguments are essentially unchanged if we talk of "O^-" instead or a linear combination of various valences.

CHARGE TRANSFER RESONANCES IN THE METALLIC STATE

Given the presence of nominally 3+ Cu ions, not only can metallic behavior occur but also a resonance will appear in the excitation spectrum due to the excitonic transition $Cu^{+3}O^{--} \to Cu^{+2}O^{-}$ at relatively low energy (several tenths of eV) on the electronic scale but large on the scale of lattice vibrations. Under certain conditions this resonance can be fairly well-defined. Then, since the metallic behavior also involves the same particle-hole transition, the quasiparticles near the Fermi surface will acquire a large renormalization. We believe this effect is the source of the high-temperature superconductivity. We shall see below that the electronic structure is propitious for a reasonable resonance.

For high-temperature superconductivity not only must this resonance be at a high energy (on the phonon scale), it must also have a large oscillator strength and be fairly localized in real space. The last condition is necessary for the quasiparticle renormalization due to the interaction with the resonance to be large due to local-field effect. Based on the considerations above, we had predicted that a charge transfer excitonic resonance (CTR) at an energy scale of about 1/2 eV must be present in the superconducting materials and its oscillator strength must be related to T_c.[13]

A resonance at about 1/2 eV with a large oscillator strength (about .75 electron per Cu^{3+} ion) has indeed been observed both in $La_{2-x}Sr_xCuO_4$ and in $YBa_2Cu_3O_7$.[28-31] In the insulating La_2CuO_4, the absorption edge is at about 2 eV, corresponding to $Cu^{++}O^{--} \rightarrow Cu^+O^-$ transitions, in accord with our considerations above.

Let us consider the excitation spectrum starting from a situation such as in Fig. 3. Keeping only one non-bonding or bonding band (β) and the anti-bonding (conduction) band (α) and neglecting any interactions, the particle-hole excitation spectrum (at some large momentum transfer) looks then as in Fig. 5a. Let us first ignore the mixing of the $\alpha\alpha$ and the $\alpha\beta$ excitations due to the Coulomb interaction and consider its effect only on the $\alpha\beta$ excitations. For large enough V, a sharp excitonic resonance with nearly zero excitation energy carrying a large fraction of all the oscillator strength of the $\alpha\beta$ transitions appears then as in Toyozawa's calculation in a different context (see Fig. 5b).[32] An important role in such a calculation , if carried out for a realistic model, will doubtless be played by the fact that r_s for the oxide metals is ≈ 3.5 Å, $i.e.$, about twice the Cu-O separation. So V is not screened by metallic processes.

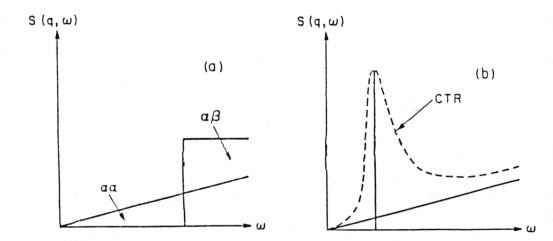

FIGURE 5 Particle-hole excitation spectrum for large momentum transfer q as a function of frequency ω (a) without, (b) including Coulomb interactions.

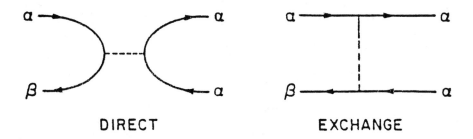

FIGURE 6 Processes leading to Landau damping of the charge transfer resonance.

An important (and hard) question is the width of such a resonance due to
i) decay into particle-hole transitions in the $\alpha\alpha$ channel and ii) scattering off the
latter. Process i) ("Landau damping") is represented diagrammatically in Fig. 6.
Note that the vertex includes direct as well as exchange scattering. The other
process ii) is due to the inherent three-body nature of the problem and described
by such diagrams as shown in Fig. 7. As is well-known from the x-ray problem and
discussed elsewhere, it cannot be estimated in conventional perturbation theory,
even for finite hole mass.[33]

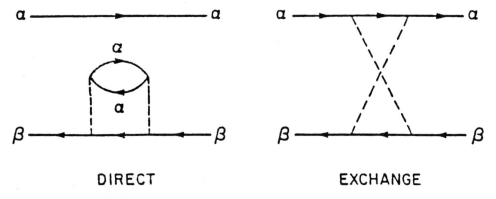

FIGURE 7 Three-body processes leading to broadening of the charge transfer resonance.

SUPERCONDUCTIVITY THROUGH CHARGE TRANSFER RESONANCES

Two kinds of charge transfer resonances should be distinguished— longitudinal and transverse. We shall see that it does not appear possible for the former which couple to electrons through their scalar potential to lead to an attractive electron-electron interaction. The latter which couple to electrons through their vector potential in general lead to an attractive electron-electron interaction which is significant only if there are large local-field effects. The transfer resonances must therefore be fairly localized in space (and in energy) and have large oscillator strengths. The electrons must also be of the tight-binding variety.

LONGITUDINAL RESONANCES

In general, the screened Coulomb interaction is given by

$$V_S = V(\epsilon_\infty - V\Pi)^{-1} = V\epsilon_L^{-1} \tag{3}$$

where Π is the irreducible polarizability and ϵ_∞ arises from the neglected (high-frequency) transitions. All objects above have four indices corresponding to the bands of incoming and outgoing particles. For the effective electron-electron interaction, we want V_{S1}, where '1' denotes $\alpha\alpha - \alpha\alpha$ particle-hole pair interactions. Denoting $\alpha\alpha - \alpha\beta$ and $\alpha\beta - \alpha\beta$ particle-hole pair interactions by '2' and '3', respectively, and for simplicity using $V_1 V_3 = V_2^2$, it is easy to show that

$$V_{S1} = \left(\frac{V_1}{\epsilon_1}\right) + \left(\frac{V_2}{\epsilon_1}\right)^2 \frac{(\Pi_{\alpha\beta,\alpha\beta} + \Pi_{\beta\alpha,\beta\alpha})}{1 - \left(\frac{V_3}{\epsilon_1}\right)(\Pi_{\alpha\beta,\alpha\beta} + \Pi_{\beta\alpha,\beta\alpha})} \tag{4}$$

where

$$\epsilon_1 = \epsilon_\infty + V_1 \Pi_{\alpha\alpha,\alpha\alpha} \tag{5}$$

is the metallic part of the longitudinal dielectric function. The charge transfer resonance (CTR) appears as a pole in the second term on the r.h.s. of (4), which may be parametrized as

$$V_{S1} = \left(\frac{V_1}{\epsilon_1}\right)\left(1 + \frac{\tilde{\omega}_0^2}{\omega^2 - \omega_T^2 - \tilde{\omega}_0^2 + i\omega\gamma}\right) \tag{6}$$

Here, $\tilde{\omega}_0^2 = \omega_0^2/\epsilon_1$ gives the screened oscillator strength and $\omega_L^2 = \omega_T^2 + \tilde{\omega}_0^2$ and γ the center frequency and width of the resonance, respectively. ω_T is the frequency of the transverse CTR. (V/ϵ_1) may be likened to the usual Coulomb parameter μ. For ω_L, ω_T large enough, there is no point in defining μ^*. An examination of (6) now reveals that it is attractive roughly only between ω_T and ω_L, which is too high a frequency range to be effective for superconductivity. At low frequencies, below ω_T, V_{S1} is always repulsive due to the self-screening of the CTR (see Fig. 8).

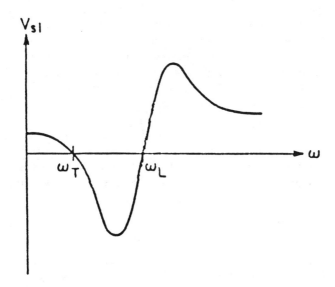

FIGURE 8 Screened Coulomb interaction as a function of frequency ω.

TRANSVERSE RESONANCES

Consider now the transverse dielectric function, which including the high-frequency contribution and the effect of the CTR may be written as

$$\epsilon_T = \epsilon_\infty - \frac{\omega_0^2}{\omega^2 - \omega_T^2 + i\omega\gamma} \tag{7}$$

Since there is no direct transverse interaction between electrons (except in QED), we need consider only the second term in (7) to obtain the induced interaction. Obviously, it is attractive for all frequencies below ω_T, the frequency of the transverse CTR. In any calculation, both the repulsive longitudinal and this attractive transverse part must be considered.

The conclusion that longitudinal excitonic resonances do not appear to lead to attractive pairing is arrived at by arguments similar to those used by Inkson and Anderson[34] in connection with proposals such as those of Little,[35] Ginzburg,[36] and Allender, Bray, and Bardeen.[37]

The transverse resonances appear, to our knowledge, not to have been suggested before, as a source of superconductive pairing. They do not have the screening problems which make longitudinal resonances unsuitable. Transverse resonances do not however couple to free electrons. Their coupling to electrons is however as important as that of longitudinal resonances for tightly bound electrons.

The general idea of pairing through electronic resonances has been discussed before.[35-37] Our ideas differ from these in some important respects. Past proposals for excitonic superconductivity have relied on there being two types of electrons, one to give band to band transitions and another to give free electrons. Moreover, only longitudinal processes have been discussed. Using this idea, Little has suggested that in $La_{2-x}Sr_xCuO_4$ superconductivity is not due to electronic resonances and that in $YBa_2Cu_3O_7$ his ideas are realized due to the bonding anti-bonding transition in the Cu-O chains.[38] According to our ideas both in $La_{2-x}Sr_xCuO_4$ and $YBa_2Cu_3O_7$ pairing is caused predominantly by the transverse charge transfer resonances. Chains are not required, nor is the presence of alternate metal and semiconductor layers. We have also presented a model in which the antiferromagnetic insulator ground state of the materials under discussion naturally arises. It is a vital part of this model that slightly away from half-filling charge transfer resonances have large oscillator strengths. This aspect is crucial to the overall understanding of these materials.

RELATIONSHIP TO EXPERIMENTS

The predicted charge transfer resonances have been observed in optical experiments at around 0.5 eV in $La_{2-x}Sr_xCuO_4$ and at around 0.7 eV in $YBa_2Cu_3O_7$.[28-31] They have half-widths of about 0.5 eV as well. They dominate the optical properties—about 2/3 of the spectral weight up to about 2 eV is contained in these resonances. They are in fact responsible for the dull black color of these materials which by the Drude term alone would have been golden.

We have also suggested experiments as a function of doping to verify the predicted correlation of the oscillator strength of the resonance with T_c.[13] These have now been done—the results are quite consistent with our ideas. For low $(x < 0.1)$ and high $(x > 0.2)$ concentration, $La_{2-x}Sr_xCuO_4$ shows little oscillator strength in the CTR. At these concentrations, these particular samples were non-superconducting. At the intermediate concentration $x \approx 0.175$, the highest T_c and the highest oscillator strength in the CTR have been observed.[29] The nonlinear dependence of T_c on this oscillator strength, to which the coupling constant λ is proportional, suggests $\lambda < 1$, i.e., weak coupling. In another set of experiments, a rough measure of the oscillator strength in various samples is found proportional to the fraction of ideal Meissner effect observed in those samples with the onset temperature always near 40 K.[30] In $YBa_2Cu_3O_7$, a high $T_c \approx 90$ K is observed with a large oscillator strength for the CTR, while in $YBa_2Cu_3O_{6.2}$, a non-superconducting compound, the CTR is absent.[31] This correlation appears to us to be a strong confirmation of the ideas presented by us.

In $YBa_2Cu_3O_7$ the average valence is 7/3 while in $YBa_2Cu_3O_{6.5}$ the Cu in the planes is in 2+ state (and that along lines is in 1+ state). There should be a general trend of correlation in the oscillator strength of the CTR with the average number of 3+ ions. This is probably loosely followed, but the non-monotonic decrease in T_c with oxygen content in this compound needs further investigation. Similarly initial addition of Sr in $La_{2-x}Sr_xCuO_4$ must lead to increased average Cu^{+++} configuration. The decrease in the oscillator strength (and T_c) with x beyond 0.2 is doubtless associated with the orthorhombic to tetragonal phase transition. We are led to suspect that at this transition the average number of Cu^{+++} decreases leading to the formation of O^- or more likely the peroxide ion $(O_2)^{--}$.

The quantitative extraction of parameters from the optical data must await experiments with polarized light on single crystals. The highly anisotropic nature of these materials leads to some difficulties in the quantitative analysis of results from experiments in polycrystals.

With such *caveat emptor*, we may proceed to extract λ from the data of Orenstein, *et al.*[28,29] with the additional and rather strong assumption that the resonances observed are almost q-independent. We then find $\lambda \lesssim 1$ for both $La_{1.82}Sr_{0.18}CuO_4$ and $YBa_2Cu_3O_7$ with $\omega_T \approx 0.5\,eV$ for the former and $\omega_T \approx 0.7\,eV$ for the latter. This yields a rather large value of T_c compared with experiments. Two things must be borne in mind—such estimates give an upper limit on λ and the usual expression for T_c is not valid for such a large range of the attractive interaction. Migdal's theorem breaks down and the corrections are doubtless in the direction of reducing T_c. This has been discussed by Grabowsky and Sham.[39] Another problem well worth looking into is the ratio Δ/T_c for $\omega_T/E_F \approx O(1)$, when λ is in the weak coupling regime. We suspect Δ/T_c will be higher than the BCS value.

With Cu in a mixed valent (2+, 3+) state and a given atom changing from one to the other configuration, it is not possible to sustain magnetic order, if, as we have taken for granted, the 3+ state in these structures is a magnetic singlet. The situation is somewhat similar to the metallic rare-earth mixed valence compounds which also do not magnetically order.

REFERENCES

1. J. G. Bednorz and K. A. Mueller, *Z. Physik B*, **64**, 189 (1986).
2. S. Uchida, H. Tagaki, K. Kitazawa, and S. Tanaka, *Jap. J. Appl. Phys. Lett.*, **26**, L1 (1987).
3. C. W. Chu, P. H. Hor, R. L. Meng, L. Gao, Z. J. Huang, and Y. Q. Wang, *Phys. Rev. Lett.*, **58**, 405 (1987).
4. R. J. Cava, R. B. Van Dover, B. Batlogg, and E. A. Rietmann, *Phys. Rev. Lett.*, **58**, 408 (1987).
5. M. K. Wu, J. R. Ashburn, C. J. Torng, P. H. Hor, R. L. Meng, L. Gao, Z. J. Huang, Y. Q. Wang, and C. W. Chu, *Phys. Rev. Lett.*, **58**, 908 (1987).
6. R. J. Cava, B. Batlogg, R. B. Van Dover, D. W. Murphy, S. Sunshine, T. Siegrist, J. P. Remeika, E. A. Rietmann, S. Zahurak, and G. P. Espinosa, *Phys. Rev.*, **58**, 1676 (1987).
7. W. Weber, *Phys. Rev. Lett.*, **58**, 1371 (1987).

8. P. Prelovsek, T. M. Rice, and F. C. Zhang, *J. Phys. C*, **20**, L229 (1987).
9. K. Miyake, S. Schmitt-Rink, and C. M. Varma, *Phys. Rev. B*, **34**, 6554 (1986).
10. D. J. Scalapino, E. Loh, and J. Hirsch, *Phys. Rev. B*, **34**, 8190 (1986).
11. P. W. Anderson, *Science*, **235**, 1196 (1987); G. Baskaran, Z. Zou, and P. W. Anderson, *Solid State Commun.*, **63**, 973 (1987) and this volume.
12. A. E. Ruckenstein, P. J. Hirschfeld, and J. Appel, *Phys. Rev. B*, **36**, 857 (1987) and this volume.
13. C. M. Varma, S. Schmitt-Rink, and E. Abrahams, *Solid State Commun.*, **62**, 681 (1987).
14. B. Batlogg, R. J. Cava, A. Jayaraman, R. B. Van Dover, G. A. Kourouklis, S. Sunshine, D. W. Murphy, L. W. Rupp, H. S. Chen, A. White, A. M. Mujsce, and E. A. Rietmann, *Phys. Rev. Lett.*, **58**, 2333 (1987).
15. L. C. Bourne, M. F. Crommie, A. Zettl, H. C. Zur Loye, S. W. Keller, K. L. Leary, A. M. Stacy, K. J. Chang, M. L. Cohen, and D. E. Morris, *Phys. Rev. Lett.*, **58**, 2337 (1987).
16. B. K. Chakraverty and J. Ranninger, *Philos. Mag. B*, **52**, 669 (1985).
17. P. Nozieres and S. Schmitt-Rink, *J. Low Temp. Phys.*, **59**, 195 (1985).
18. P. Ganguly and C. N. R. Rao, *J. Solid State Chem.*, **53**, 198 (1984).
19. D. Vaknin, S. K. Sinha, D. E. Moncton, D. C. Johnston, J. Newsam, C. R. Safinya, and H. E. King, Jr., *Phys. Rev. Lett.*, **58** , 2802 (1987).
20. S. Schmitt-Rink, K. Miyake, and C. M. Varma, *Phys. Rev. Lett.*, **57**, 2575 (1986).
21. D. R. Harshman, G. Aeppli, B. Batlogg, R. J. Cava, E. J. Ansaldo, J. H. Brewer, W. Hardy, S. R. Kreitzman, G. M. Luke, D. R. Noakes, and M. Senba, *Phys. Rev. B*, in press.
22. B. Golding, N. O. Birge, W. H. Haemmerle, R. J. Cava, and E. Rietmann, preprint.
23. E. Abrahams and C. M. Varma, preprint.
24. L. F. Mattheiss, *Phys. Rev. Lett.*, **58**, 1028 (1987).
25. L. F. Mattheiss and D. R. Hamann, *Solid State Commun.*, in press.
26. J. Zaanen, G. A. Sawatzky, and J. W. Allen, *Phys. Rev. Lett.*, **55**, 418 (1985).
27. G. A. Sawatzky and J. W. Allen, *Phys. Rev. Lett.*, **53**, 2339 (1984).
28. J. Orenstein, G. A. Thomas, D. H. Rapkine, C. G. Bethea, B. F. Levine, R. J. Cava, E. A. Rietmann, and D. W. Johnson, Jr., *Phys. Rev. B*, in press.
29. J. Orenstein, G. A. Thomas, D. H. Rapkine, C. G. Bethea, B. F. Levin, B. Batlogg, R. J Cava, D. W. Johnson, Jr., and E. A. Rietmann, preprint.
30. S. Etemad, D. E. Aspnes, M. K. Kelly, R. Thompson, J. M. Tarascon, and G. W. Hull, preprint.
31. K. Kamaras, C. D. Porter, M. G. Doss, S. L. Herr, D. B. Tanner, D. A. Bonn, J. E. Greedan, A. H. O'Reilly, C. V. Stager, and T. Timusk, preprint.
32. Y. Toyozawa, M. Inoue, T. Inui, M. Okazaki, and E. Hanamura, *J. Phys. Soc. Japan*, **21**, 208 (1966); *ibid.*, 209 (1966).
33. A. E. Ruckenstein and S. Schmitt-Rink, *Phys. Rev. B* , **35**, 7551 (1987).
34. J. C. Inkson and P. W. Anderson, *Phys. Rev. B*, **8**, 4429 (1973).

35. W. A. Little, *Phys. Rev.*, **134**, A1416 (1964).
36. V. L. Ginzburg, *Usp. Fiz. Nauk*, **101**, 185 (1970) [*Sov. Phys.-Usp.*, **13**, 335 (1970)].
37. D. Allender, J. Bray, and J. Bardeen, *Phys. Rev. B*, **7**, 1020 (1973); *Phys. Rev. B*, **8**, 4433 (1973).
38. J. P. Collman, J. T. McDevitt, and W. A. Little, preprint.
39. M. Grabowsky and L. J. Sham, *Phys. Rev. B*, **29**, 6132 (1984).

V. J. Emery
Physics Department
Brookhaven National Laboratory
Upton, New York 11973

Theory of High-T_c Superconductivity in Oxides[†]

It is shown that the properties of high-T_c oxide superconductors are consistent with a model in which the charge carriers are holes in the O($2p$) states and the pairing is mediated by strong coupling to local spin configurations on the Cu sites.

The discovery that a number of superconducting oxides have remarkably high transition temperatures[1-3] has revived the discussion of unconventional pairing mechanisms in solids. Although phonon exchange[4] seems capable of producing transition temperatures above 30 K, as required[1] for doped La_2CuO_4, it is more difficult to imagine that it can be responsible for superconductivity above 90 K as attained[3] in $YBa_2Cu_3O_{9-\delta}$ and related materials. In this letter, it will be shown that an analysis of the currently known properties of the oxides leads naturally to an alternative mechanism—strong coupling to local spin configurations—which can readily produce transition temperatures of the required magnitude. In some respects the mechanism resembles anisotropic pairing produced by exchange of spin fluctuations, proposed for organic superconductors[5] and heavy-fermion systems.[6]

[†]Reprinted from *Phys. Rev. Lett.*, **58**, 2794 (1987).

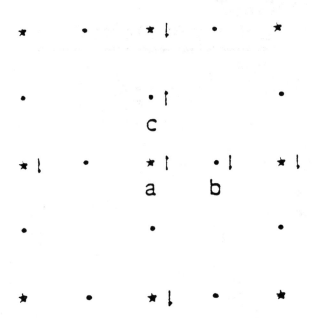

FIGURE 1 The structure of the CuO$_2$ planes. The stars indicate Cu sites and the dots oxygen sites. The particular configuration of spins is discussed in the text.

But the electronic and crystal structure of the oxides leads to a much stronger effect. The picture is quite different from Anderson's resonant-valence-bond model[7] and has an insulating limit which is antiferromagnetic (AF) rather than dimerized.

A common feature of the high-T_c oxides is the quasi two-dimensional motion of electrons within CuO$_2$ planes[8-11] which have a structure[12] shown in Fig. 1. It will be assumed that the Hamiltonian for a single plane is given by an extended Hubbard model,

$$H = \sum_{\mathbf{i,j},\sigma} \epsilon_{\mathbf{ij}} a^\dagger_{\mathbf{i}\sigma} a_{\mathbf{j}\sigma} + \frac{1}{2} \sum_{\substack{\mathbf{i,j} \\ \sigma,\sigma'}} U_{\mathbf{ij}} a^\dagger_{\mathbf{i}\sigma} a_{\mathbf{i}\sigma} a^\dagger_{\mathbf{j}\sigma'} a_{\mathbf{j}\sigma'} \ . \tag{1}$$

Here, $\mathbf{i}$ labels a copper or an oxygen site, and the vacuum consists of Cu$^+$ (all d states occupied) and O^{2-} (all p states occupied). The operators $a^\dagger_{\mathbf{i}\sigma}$ create holes with spin σ in the Cu($3d_{x^2-y^2}$), O($2p_x$), or O($2p_y$) states which are the ones most strongly hybridized by overlap integrals.[8-10] The site-diagonal terms ($\epsilon_{\mathbf{ii}}, U_{\mathbf{ii}}$) are ($\epsilon_p, U_p$) and ($\epsilon_d, U_d$) for O($2p$) and Cu($3d$) states, respectively, and the only other nonvanishing terms are hopping integrals $\epsilon_{\mathbf{ij}} = \pm t$ and an interaction $U_{\mathbf{ij}} = V$

between neighboring sites (Cu and O). The parameters are not well known but, to fix ideas, it will be assumed that $t = 1.3$–1.5 eV, $\epsilon \equiv \epsilon_p - \epsilon_d = 1$ eV, $U_p = 2$–3 eV, $U_d = 5$–6 eV, and $V = 1$–2 eV. ($\epsilon > 0$ since ϵ_p and ϵ_d are hole energies.) These values are consistent with Mattheiss's effective tight-binding model,[9] regarded as equivalent to applying mean-field theory to H. [The mean-field hopping integral has a contribution from V, and site energies are equal if $\epsilon = (2U_d - U_p)/8$.]

Much of the discussion will be concerned with doped La$_2$CuO$_4$, whose properties are best known at the present time—application of the model to the higher-T_c materials will be considered at the end. The number of holes per Cu site will be denoted by $1 \pm \delta$, where δ is determined by doping, oxygen defects, and the states of atoms outside the CuO$_2$ planes. According to electronic band-structure calculations,[9,10] La$_2$CuO$_4$ has a half-filled band ($\delta = 0$) and an almost perfectly nested Fermi surface—which is quite difficult to reconcile with the observation that La$_2$CuO$_4$ is a superconductor.[13] A possible explanation is that, contrary to Refs. 9 and 10, the La($5d$) band actually dips below the Fermi level, thereby removing electrons from the CuO$_2$ planes and making $\delta > 0$ in undoped stoichiometric La$_2$CuO$_4$. Then doping with divalent elements such as Sr or Ba would remove electrons from the La($5d$) and and increase δ, whereas oxygen defects would have the opposite effect. This view is consistent with studies of the La$_4$L$_3$ edge[14] which revealed that doping produced a systematic increase in the density of unoccupied states having d symmetry with respect to the La site. The strong pressure dependence[2] of T_c might also be produced by variations in the occupancy of the La($5d$) band.

For $t = 0$ and $\delta = 0$, the ground state of H will have exactly one hole on each Cu site (Cu^{2+}). Hopping may be included by the elimination of the oxygen sites to give an effective Hamiltonian for motion on the Cu. To second order this is a simple Hubbard model with hopping integral $t_d = t^2/(\epsilon + V)$. Hirsch[15] has carried out an extensive Monte Carlo study of this model: The ground state is an AF insulator, and when the weak hopping between CuO$_2$ planes is taken into account there will be long-range AF order at finite temperature. Recently, it has been established[16,17] that there is indeed AF order in La$_2$CuO$_4$ samples which presumably have δ close to zero.

There is a gap Δ between the energies of occupied and unoccupied states of the Hubbard model for a half-filled band.[15] Consequently, any additional holes will go into the O($2p$) states if the site energy lies inside the gap. This will be the case for $t = 0$ if $U_d > \epsilon + 2V$, and it will persist for a finite range of values of t since $\Delta \approx U_d$ for intermediate coupling. Experiments on doped La$_2$CuO$_4$ fit well with this picture and it will be assumed to hold true. The most direct evidence comes from an x-ray-absorption near-edge study[14] of La$_{2-x}$Ba$_x$CuO$_4$ and La$_{2-x}$Sr$_x$CuO$_4$ which found that the copper remained in the Cu^{2+} state for all x in the range $0 \leq x \leq 0.3$.

An effective Hamiltonian for the O($2p$) holes may be obtained by elimination of the available Cu($3d$) states. To second order in t the kinetic energy is given by

$$H_0 = \sum_{\mathbf{k},\sigma} t^2 e_{\mathbf{k}} G_{\mathbf{k}}(\epsilon + 2V - \mu) b_{\mathbf{k}\sigma}^{\dagger} b_{\mathbf{k}\sigma} \ , \tag{2}$$

where $G_k(\omega)$ is the one-particle Green's function and μ the chemical potential of the Cu(3d) holes,

$$e_{\mathbf{k}} = 2(2 - \cos k_x a - \cos k_y a), \tag{3}$$

where a is the lattice spacing, and $b_{\mathbf{k}\sigma}$ is the Fourier transform of $a_{\mathbf{j}\sigma}$ for *all oxygen* states, divided by $e_{\mathbf{k}}^{1/2}$ for normalization. The energy spectrum is determined by the shape of $G_{\mathbf{k}}$: Even for an insulator, $G_{\mathbf{k}}$ may vary rapidly but without discontinuity in the neighborhood of the ideal Fermi surface S_F, on a scale determined by the mean free path or localization length. As a result, $e_k G_k$ would have a minimum just inside S_F. From Hirsch's Monte Carlo calculations,[15] it may be inferred that this picture is valid for intermediate coupling but, for very large U_d, $G_{\mathbf{k}}$ depends weakly on $\mathbf{k}$ and the minimum in the spectrum moves to the zone corners. Experiment supports the intermediate-coupling picture: The plasma frequency ω_p in doped La$_2$CuO$_4$ obtained directly[18] and from the London penetration depth[19] ($\lambda = c/\omega_p$, where c is the velocity of light) varies slowly with δ, which requires[18] that the Fermi surface be close to S_F. The values of $\hbar\omega_p$ (0.8 eV) and the density of states[20] ($\gamma = 6$ mJ/mole-Cu K^2) may both be fitted with a spectrum $-\bar{t}e_{\mathbf{k}}$ with $\bar{t} = 0.13$ eV, which is much smaller than the band-structure value[9] (1.6 eV) but is compatible with Eqs. (2) and (3) since t^2/Δ is about 0.5 eV and the density of states is increased by the factor $G_{\mathbf{k}}(\epsilon + 2V - \mu)$. (There is also mass enhancement due to the interactions.) Thus, it is reasonable to have a picture in which the charge carriers, mainly on the oxygen atoms, have a narrow band $-\bar{t}e_{\mathbf{k}}$ and a number density $n_c = n\delta$, where $n = 10^{22}$ cm^{-3} is the number density of the copper.

Elimination of the Cu sites will also produce an effective attractive coupling between the O(2p) holes which is responsible for superconductivity. The essential point, and the reason for high T_c, is that the coupling is strong because O(2p)-Cu(3d) exchange interactions are larger than Cu(3d)-Cu(3d) exchange. (This is why AF order is destroyed for relatively small δ.) Consider the situation shown in Fig. 1, where there are O(2p) holes of opposite spins at b and c. There is an instantaneous magnetic moment at a, which is somewhat smaller than the full moment since the spins are delocalized. It is opposite to the spin of one of the other of the O(2p) holes. Exchange of the holes at b and c may be accomplished by an (ab) interchange, followed by an (ac) interchange. An estimate of the associated energy is $-v_0$ where

$$v_0 = \langle (n_{d\uparrow} - n_{d\downarrow})^2 \rangle \left(\frac{t^2}{U_p + \epsilon} + \frac{t^2}{U_d - \epsilon} \right)^2 (2J)^{-1}. \tag{4}$$

Here, the quantity in brackets is the matrix element for the (ab) or (bc) exchange. Delocalization of the Cu moment has been taken into account by the omission of V from the denominators and by inclusion of the factor $\langle (n_{d\uparrow} - n_{d\downarrow})^2 \rangle$, where $n_{d\sigma}$ is the number operator for a Cu hole of spin σ at the site.[15] This factor is about 0.73 for $U_d = 4t_d$. The denominator $2J$ in Eq. (4) is the energy to break the four bonds between a and its Cu neighbors. For large U_d, $J = 4t_d^2/U_d$ but, for weaker coupling, this value is reduced because the probability that a is surrounded by opposite spins is less than 1^{15} and because the holes at b and c partially block the exchange. The interaction v_0 is the primary source of pairing, and the BCS transition temperature[21] is given by the condition for a nontrivial solution of

$$\Delta_{\mathbf{k}} = -\frac{1}{(2\pi)^2} \int d\mathbf{k}' v_{\mathbf{k}-\mathbf{k}'} \frac{\tanh\left(\frac{1}{2}\beta_c \epsilon_{\mathbf{k}'}\right)}{2\epsilon_{\mathbf{k}'}} \Delta_{\mathbf{k}'}, \tag{5}$$

where $\beta_c = (k_B T_c)^{-1}$, and $\epsilon_{\mathbf{k}} = -\bar{t} e_{\mathbf{k}}$. This is a two-dimensional mean-field-theory result and T_c would be somewhat reduced by phase fluctuations. The potential $v_{\mathbf{q}}$ is given by

$$v_{\mathbf{q}} = U_p - 0.57 v_0 (\cos q_x a + \cos q_y a), \tag{6}$$

where the factor 0.57 allows for six neighbors coupled by exchange and for the transformation from $a_{i\sigma}$ to $b_{\mathbf{k}}$, averaged over the Fermi surface for $\delta = 0.15$. (A second-neighbor coupling generated by this transformation has been omitted.) The cosine potential (6) is separable and Eq. (5) may be solved analytically. The solution which is even in $\mathbf{k}$ and excludes the repulsion U_p is the d state

$$\Delta_{\mathbf{k}} = \Delta_0 (\cos k_x a - \cos k_y a) \tag{7}$$

which gives

$$k_B T_c \sim E_0 e^{-7\pi\bar{t}/v_0}, \tag{8}$$

where the prefactor is a cutoff of the order of the O(2p) Fermi energy (600 K). All of this assumes $\delta = 0.15$. The value of T_c is sensitive to the assumed parameters but it is easy to obtain transition temperatures between 30 and 40 K since v_0 is of the order of 1 eV. It is characteristic of the oxide superconductors that $E_0/k_B T_c$ is relatively small,[22] in this case about 20.

As δ increases, the change in the Fermi surface will cause both E_0 and the magnitude of the exponent in Eq. (8) to increase, and eventually T_c will fall. This may not be the reason for the absence of superconductivity in doped La$_2$CuO$_4$ for $x > 0.2$, which might have more to do with the role of oxygen defects.[23] Indeed, the CuO$_2$ planes in YBa$_2$Cu$_3$O$_{9-\delta}$ are quite similar to those in La$_2$CuO$_4$, and the high T_c may simply be a consequence of increased carrier density, as expected from the present model. For some values of the parameters, the interchange of a and b lowers the energy, and a real bound pair (spin polaron) will form. This does not appear to happen in doped La$_2$CuO$_4$ but it should be kept in mind in the interpretation of experiments on other materials.

It is important to notice that in obtaining superconductivity from purely repulsive interactions, the binding of a pair was a consequence of their coupling to other degrees of freedom, just as for the exchange of spin fluctuations or other collective modes. This is essential. In the strong-coupling limit, it is possible to rearrange a repulsive *pairing force* to give an effective attractive interaction[24] of order t^2/U, without involving other electrons, but it can be shown that this cannot give rise to superconductivity.

ACKNOWLEDGEMENTS

I have benefitted from many discussions with Dr. J. Davenport, particularly conserning questions of electronic structure. I am also grateful to G. Aeppli, J. D. Axe, G. Collin, R. Comés, D. Cox, A. Goldman, D. Moncton, G. Shirane, T. Uemura, and particularly J. Tranquada for discussions of experiments.

This work was supported by the Division of Materials Sciences, U. S. Department of Energy, under Contract No. DE-AC02-76CH00016.

Note added.—A recent muon-spin-relaxation measurement[25] shows that ω_p is 1.5 to 2 times as high in $Ba_2YCu_3O_{9-\delta}$ (with $\delta = 2.1$) as in doped La_2CuO_4, and also gives evidence for s-wave pairing. The change in ω_p suggests that the Fermi surface for $O(2p)$ holes may be at the zone corners, where in fact d-state pairing is less favorable than s-state pairing. In the present model, it is possible to construct a uniform s-state wave function which does not allow two $O(2p)$ holes to occupy the same site,[26] since there are two oxygen sites per unit cell. It may also be that U_p, which only enters into the gap Eq. (5) and does not affect the preceding argument, is small. The experiment could also be explained by the existence of real pairs rather than Cooper pairs. Although Cooper pairing was assumed in Eqs. (5)–(7), since it seems to be consistent with other experiments, real pairs should occur[26] in the present model for small enough δ.

REFERENCES

1. J. G. Bednorz and K. A. Müller, *Z. Phys. B*, **64**, 189 (1986).
2. S. Uchida, H. Takagi, K. Kitazawa, and S. Tanaka, *Jpn. J. Appl. Phys. Part 2*, **26**, L1 (1987); C. W. Chu, P. H. Hor, R. L. Meng, L. Gao, Z. J. Huang, and Y. Q. Wang, *Phys. Rev. Lett.*, **58** 405 (1987); R. J. Cava, R. B. van Dover, B. Batlogg, and E. A. Rietman, *Phys. Rev. Lett.*, **58**, 408 (1987).
3. M. K. Wu, J. R. Ashburn, C. J. Torng, P. H. Hor, R. L. Meng, L. Gao, Z. H. Huang, Y. Q. Wang, and C. W. Chu, *Phys. Rev. Lett.*, **58**, 908–10 (1987); R. J. Cava, B. Batlogg, R. B. van Dover, D. W. Murphy, S. Sunshine, T. Siegrist, J. P. Remeika, E. A. Rietman, S. Zahurak, and G. P. Espinosa, *Phys. Rev. Lett.*, **58**, 1676 (1987).
4. W. Weber, *Phys. Rev. Lett.*, **58**, 1371, 2154(E) (1987).
5. V. J. Emery, *J. Phys. (Paris) Colloq.*, **44**, C3-977 (1983), and *Synth. Met.*, **13**, 21 (1986).
6. D. J. Scalapino, E. Loh, Jr., and J. E. Hirsch, *Phys. Rev. B*, **34**, 8190 (1986); K. Miyake, S. Schmitt-Rink, and C. M. Varma, *Phys. Rev. B*, **34**, 6554 (1986).
7. P. W. Anderson, *Science*, **235**, 1196 (1987).

8. J.D. Jorgensen, H. B. Schüttler, D. G. Hinks, D. W. Capon II, D. Zhang, M. B. Brodsky, and D. J. Scalapino, *Phys. Rev. Lett.*, **58**, 1024 (1987).

9. L. F. Mattheiss, *Phys. Rev. Lett.*, **58**, 1028 (1987).

10. J. Yu, A. J. Freeman, and J.-H. Xu, *Phys. Rev. Lett.*, **58**, 1035 (1987).

11. T. Siegrist, S. Sunshine, D. W. Murphy, R. J. Cava, and S. M. Zahurak, *Phys. Rev. B*, **35**, 7137 (1987).

12. Possible lattice distortions are ignored since they do not play an essential role in the present argument.

13. J. Beille, R. Cabanel, C. Chaillout, B. Chevallier, G. Demazeau, F. Deslandes, J. Etourneau, P. Lejay, C. Michel, J. Provost, B. Raveau, A. Sulpice, J.-L. Tholence, and R. Tournier, *C. R. Acad. Sci. Ser. 2*, **304**, 1097 (1987); G. Collin and D. Jérome, unpublished.

14. J. M. Tranquada, S. M. Heald, A. R. Moodenbaugh, and M. Suenaga, *Phys. Rev. B*, **35**, 7187 (1987). A more detailed presentation of this work, to be published, gives a critical discussion of related experiments which draw different conclusions.

15. J. Hirsch, *Phys. Rev. B*, **31**, 4403 (1985).

16. R. L. Greene, H. Maletta, T. S. Plaskett, J. G. Bednorz, and K. A. Müller, to be published.

17. D. Vaknin, S. K. Sinha, D. E. Moncton, D. C. Johnston, J. M. Newsam, C. R. Safinya, and H. E. King, Jr., *Phys. Rev. Lett.*, **58**, 2802 (1987); T. Freltoft, J. P. Remeika, D. E. Moncton, A. S. Cooper, J. E. Fischer, P. Harshman, G. Shirane, S. K. Sinha, and D. Vaknin, *Phys. Rev. B*, to be published.

18. S. Tajima, S. Uchida, S. Tanaka, S. Kanbe, K. Kitazawa, and K. Feuki, *Jpn. J. Appl. Phys. Part 2*, **26**, L432 (1987).

19. G. Aeppli, R. J. Cava, E. J. Ansaldo, J. H. Brewer, S. R. Kreetzman, G. M. Luke, D. R. Noakes, and R. F. Kiefl, *Phys. Rev. B*, **35**, 7129 (1987); W. J. Kossler, J. R. Kempton, X. H. Yu, H. E. Schone, Y. J. Uemura, A. R. Moodenbaugh, M. Suenaga, and C. E. Stronach, *Phys. Rev. B*, **35**, 7133 (1987).

20. B. Batlogg, A. P. Raminez, R. J. Cava, R. B. van Dover, and E. A. Rietman, *Phys. Rev. B*, **35**, 5340 (1987).

21. J. Bardeen, L. N. Cooper,. and J. R. Schrieffer, *Phys. Rev.*, **108**, 1175 (1975).

22. C. Noguera and P. Garoche, (to be published), have reached this conclusion by analyzing different experiments.

23. G. Collin, private communication.

24. J. E. Hirsch, *Phys. Rev. Lett.*, **54**, 1317 (1985).

25. D. R. Harshman, G. Aeppli, E. J. Ansaldo, B. Batlogg, J. H. Brewer, J. F. Carolan, R. J. Cava, M. Celio, A. C. D. Chaklader, W. N. Hardy, S. R. Kreitzman, G. M. Luke, D. R. Noakes, and M. Senba, to be published.

26. V. J. Emery, in *Proceedings of the NATO Advanced Research Workshop on Organic and Inorganic Low-dimensional Crystalline Materials*, Minorca, Spain, 3–8 May 1987, to be published.

P. A. Lee, G. Kotliar, and N. Read
Department of Physics
Massachusetts Institute of Technology
Cambridge, Massachusetts 02139

Fermi Liquid Theory of La$_{2-x}$ Sr$_x$CuO$_4$: Optical Properties†

We treat a model with orbitals on the copper and oxygen sites and hopping between them. In the limit of infinite U on the Cu site, the problem is treated using the slave boson method. In the presence of doping, we find a Fermi liquid solution with mass enhancement proportional to x^{-1}. A new feature in the optical absorption is found with oscillator strength proportional to x, which can explain a peak at $0.5\,$eV observed experimentally. The interaction between quasi-particles via the exchange of slave bosons will be discussed.

The physics of the high T_c superconductor La$_{2-x}$Sr$_x$CuO$_4$ is often discussed in terms of the Hubbard model. A number of authors have pointed out that it is important to include the copper and oxygen orbitals in terms of understanding the charge transfer excitation[1] and possibly even the mechanism for superconductivity.[2,3] We consider a model in which $d_{i,\sigma}^+$ creates a hole in the Cu($3d_{x^2-y^2}$) orbital at energy ε_d^o, $c_{i\sigma}^+$ creates a hole in the O($2p_x$ or $2p_y$) orbital at energy ε_p and a hopping matrix element t_{pd} connects the nearest Cu and O neighbors. We introduce a Hubbard U_d and U_c on the copper and oxygen sites, but we shall treat the limits $U_d = \infty$ and

†Reprinted from *(Proc. of Yamada Conference)*, *Physica*, **148**B, 274-244, (1987).

$U_c = 0$. U_d has been estimated to be roughly $6\,\text{eV}$ [4] which is much larger than any other energy scale in the problem, so that $U_d = \infty$ (absence of Cu^{3+}) is a reasonable starting point. We also assume that $\varepsilon_p - \varepsilon_d^o > 0$ so that the undoped La_2CuO_4 nominally consists of Cu^{2+} and O^{2-} and the holes introduced by doping with Sr occupies of the oxygen orbitals. As we shall see, $\varepsilon_p - \varepsilon_d^o \equiv D$ plays the role of U in the Hubbard model, and this is consistent with the analysis of the trends in transition metal oxides.[5] For x small, U_c does not interfere with band formation because the probability of double occupation is small, so that setting $U_c = 0$ would not change the physics qualitatively. We should keep in mind that U_c enters into the estimate of the super-exchange J between neighboring copper spins, because that involves double occupation of the oxygen site in between, so that $J = 4t_{pd}^4(\varepsilon_p - \varepsilon_d^o)^{-2}(\varepsilon_p - \varepsilon_d^o + U_c)^{-1}$.

The $U_d = \infty$, $U_c = 0$ limit can be treated by introducing a slave boson b_i on each Cu site. The problem is further simplified by noting that only a single combination of the two oxygen orbitals in the unit cell couples to the Cu, so that we have a two-band model. For $U_d = 0$, we have the familiar band structure

$$E_{1,[2]}(k) = \frac{1}{2}(\varepsilon_p + \varepsilon_d^o - [+]R_k) \tag{1}$$

where $R_k = ((\varepsilon_p - \varepsilon_d^o)^2 + 16t_{pd}^2\gamma_k^2)^{1/2}$ and $\gamma_k^2 = \sin^2(k_x/2) + \sin^2(k_y/2)$. For one hole per unit cell, the lower band E_1 is half filled.

The present model is formally identical to the Anderson lattice and we can take over earlier work which treats the problem in a large orbital degeneracy (N) expansion.[6,7] We adopt a convention where the hopping matrix element is denoted by $t/N^{1/2}$, (so that when $N = 2, t = \sqrt{2}t_{pd}$). We assume that the N-fold degenerate d states can accommodate $Q = Nq_O$ holes and we set $q_O = 1/2$. The deviation from half-filling δ is defined by writing the total number of holes per unit cell as $H = Q(1 + \delta)$. We enforce the constraint

$$b_i^+ b_i + d_i^+ d_i = Q. \tag{2}$$

We consider the large N limit only as a formal device to produce a consistent loop expansion, and we set $N = 2$ for spin degeneracy in our final results. It is only when $N = 2$ and $Q = 1$ that the constraint Eq. (2) corresponds to the infinite U_d model.

To lowest order in N^{-1}, we have mean field theory, where $\langle b_i \rangle = b_0 \equiv \sqrt{N}r_o$, and the effective hopping matrix element becomes $\sigma_O = (t/\sqrt{N})b_0 = tr_O$. At the same time the position of the d level is renormalized from ε_d^o to ε_d so that the renormalized band structure $E_{1[2]}$ and R_k are given by Eq. (1) with $t_{pd} \to \sigma_O$ and $\varepsilon_d^o \to \varepsilon_d$. These bands describe the quasi-particles and the chemical potential is determined by filling the lower band by H/N holes. The Fermi surface then contains H holes and Luttinger's theorem is obeyed. The mean-field parameters r_o and ε_d are given by the equations

$$r_o^2 + \sum_k u_k^2 f(E_1(k)) = q_o \tag{3}$$

$$(\varepsilon_d - \varepsilon_d^o) = (t/r_o) \sum_k u_k v_k \gamma_k f(E_1(k)) \tag{4}$$

where f is the Fermi factor and $u_k^2[v_k^2] = 1/2(1 + [-](\varepsilon_p - \varepsilon_d)/R_k)$ is the $d[c]$ weight in the E_1 band. Eq. (3) is just the constraint Eq. (2) in mean field. We have solved Eq. (3) and (4) numerically, but we can learn much by examining the limit $4t \ll D$ where an analytic solution is possible. Our model contains two dimensionless parameters D/t and δ. For $\delta < 0$, the shift in ε_d can be calculated perturbatively from Eq. (4) and is small, of order t^2/D. Using $q_o = H/N - q_o\delta$, Eq. (3) can be written as

$$r_o^2 = -\frac{\delta}{2} + \sum_k (1 - u_k^2) f(E_1(k)) \tag{5}$$

For $t \ll D$, $u_k^2 \approx 1$ and we obtain $r_o^2 = -\delta/2$. The bandwidth of the two bands are given by $8r_o^2 t^2/(\varepsilon_p - \varepsilon_d)$ and we see that they are narrowed by the factor $|\delta|$ compared with the bare bandwidth $8t_{pd}^2/D$. This band narrowing is familiar in the Hubbard model in the large U/t limit. It represents the physical picture that in a Fermi-liquid picture which satisfies Luttinger's theorem, we represent δ holes in the lower Hubbard band by $H = N/2(1 + \delta)$ holes which must acquire a larger effective mass $m^*/m \approx \delta^{-1}$.

We can see from Eq. (4) that $r_o = 0$ at $\delta = 0$ and the perturbative solution we found disappears for $\delta > 0$. It turns out that another solution can be found where ε_d is renormalized by a large amount until it is close to ε_p, i.e., $\varepsilon_d - \varepsilon_p = (4t^2/D) \sum_k \gamma_k^2 f(E_1)$. Provided $\sigma_o \ll \varepsilon_d - \varepsilon_p$ (and this will be shown to be consistent), the term $\sum(1 - u_k^2) f$ in Eq. (5) can be expanded to give $r_o^2 4t^2/(\varepsilon_d - \varepsilon_p)^2 \sum_k \gamma_k^2 f(E_1)$ which is much greater than r_o^2. Combined with Eq. (4), Eq. (5) now has the solution $r_o^2 \approx (\delta/2)(2t/D)^2 \sum_k \gamma_k^2 f(E_1)$. Putting these solutions into the band structure, we find that the bandwidths are still given by $8\delta t_{pd}^2/D$ as in the $\delta < 0$ case. The picture, as shown in Fig. 1, is that as δ changes signs, the lower band shifts discontinuously from near ε_d to $(4t^2/D) \sum_k \gamma_k^2 f(E_1)$ below ε_p. The jump in ε_d is dictated by the physical requirement that the chemical potential must be near ε_p for $\delta > 0$, because the additional holes are expected to occupy oxygen sites. Indeed, the oxygen hole can delocalize by virtual hopping onto the copper sites and the energy gained is just $\approx 4t_{pd}^2/D$ from perturbation theory, thus explaining the location of the narrow band below ε_p.

Physically we do not expect the band narrowing to continue for arbitrarily small $|\delta|$, and we expect a Néel ground state at and near half filling. Thus we should compare J with the delocalization energy $4\delta t_{pd}^2/D$ and we expect that the above Fermi liquid picture will be a good starting point for $\delta > t_{pd}^2/D(D + U_c)$.

We now focus on $\delta > 0$, since $\delta = x$ for Sr doping. Our mean-field solution predicts an interesting new structure in the conductivity $\sigma(\omega)$ corresponding to transition between bands E_1 and E_2. The conductivity can be computed using the Kubo formula in terms of the current operator

$$j = i \sum_{jk} t_{jk}(b_j d_j^+ c_k - \text{h.c.}) \tag{6}$$

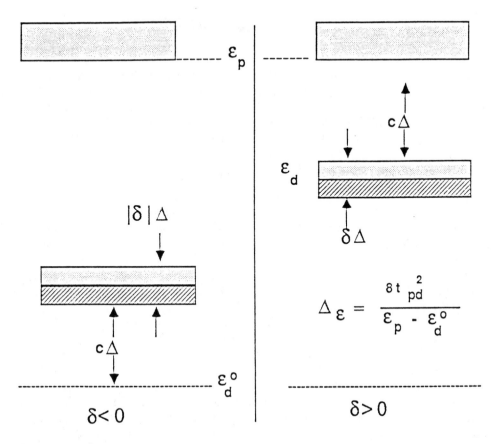

FIGURE 1 Schematic illustration of the renormalized hole bands. Shaded area indicates filled states. The constant $C = \sum_k \gamma_k^2 f(E_1(k))$ is of order unity.

where t_{jk} is the nearest neighbor hopping matrix element t_{pd}. For low frequency excitation we replace b_j with b_o and $\sigma(\omega)$ can be computed using the renormalized band structure. We find a Drude term at $\omega = 0$ corresponds to particle-hole excitation in the lower band and an excitation with a gap $\approx \varepsilon_p - \varepsilon_d \approx (4t^2/D) \sum_k \gamma_k^2 f(E_1)$ for $4t \ll D$ corresponding to interband absorption. These two features have approximately equal oscillation strength of $e^2 n \delta 4t^2/D k_F^2$ where n is the density of holes per volume (approximately 1 per Cu). It is convenient to introduce a band mass m_b by $k_F^2/m_b = 4t^2/D$ which correspond to the unrenormalized bandwidth. Then the oscillation strength can be written as $(e^2 n/m_b)\delta$. On the other hand, for high frequency excitations we must use the boson propagation $\int \langle b^+(t)b(o)\rangle e^{i\omega t}$ which has a pole at $\omega \approx \varepsilon_d - \varepsilon_d$ in computing $\sigma(\omega) \approx \int \langle j(t)j(o)\rangle e^{i\omega t} dt$. This gives rise to absorption at $\varepsilon_p - \varepsilon_d$ with oscillation strength $\approx (e^2 n/m_b)(1-\delta)$. These transitions exhaust the expected oscillation strength in this model. To understand physically the weight δ associated with the optical transition, we note that the excitation in the E_1 bands consist mainly of Cu states with a small admixture proportional to δ of O-states. Since for $t \ll D$ the charge on the Cu is fixed, the Cu states contribute only spin, and the oxygen states spin and charge (both of order δ), to the excitation. Thus, the optical transition involves only charge transfer between O sites, with an effective matrix element via the Cu site.

Recent optical reflectivity experiments[8,9] on polycrystalline samples have revealed a peak at 2 eV in the undoped as well as the doped sample. We would interpret this as $\varepsilon_p - \varepsilon_d$. An additional peak emerges at 0.5 eV upon doping, with oscillation strength linear in doping x. It is natural for us to interpret this peak as the interband transition, so the $\varepsilon_p - \varepsilon_d \approx 0.5$ eV. We note that within our model, this feature is a signature of a Fermi liquid and is not necessarily directly related to superconductivity. Indeed, in our model the contribution of this peak to the static dielectric constant ε is of order δ. Estimates based on the reported oscillator strength[8] also show a contribution to ε that is $\leq$ unity. In our opinion, this casts doubts on the interpretation that the 0.5 eV structure is the excitation responsible for an excitonic mechanism of superconductivity.[10]

We have followed Refs. (6,7) and extended the theory to first order in $1/N$, *i.e.*, including boson fluctuations. Just as in the Anderson lattice, we found that the compressibility $dn/d\mu$ is unrenormalized, *i.e.*, $dn/d\mu \approx \rho_0$ where ρ_0 is the bare density of states corresponding to the bare bandwidth $8t_{pd}^2/D$. In terms of Fermi-liquid theory,

$$\frac{dn}{d\mu} = \rho/(1 + F_o^s)$$

where $\rho \approx \delta^{-1}\rho_0$ is the renormalized density of states, which implies that $F_o^s \approx \delta^{-1} \gg 1$. The magnetic susceptibility χ is proportional to ρ with $F_o^a \approx O(1/N)$. Again, we expect χ to saturate at J^{-1} when $4\delta t_{pd}^2/D < J$.

We also computed the effective interaction between two quasi-particles by the exchange of a single boson. We find that in the static limit $\omega = 0$, $V_{\text{eff}}(q = 0)\rho = 1$. The full q dependence requires some numerical work which is in progress, but our preliminary estimates indicate that contrary to the Anderson lattice space, where single boson exchange leads to a very weak d-wave pairing,[11] a similar pairing does not occur here. Essentially, the single boson exchange describes a hard-core repulsion due to the infinite U_d constraint.

By going to next order in $1/N$ and including the exchange of two bosons, the exchange interaction J is recovered between quasi-particles. As noted earlier, the quasi-particles at the Fermi level are mostly spin excitations on the copper, and so they retain the exchange interaction one deduces for the local moment on the copper site via perturbation theory. When these exchange interactions are taken into account, our theory looks very much like the mean-field treatment of the effective Hamiltonian for the Hubbard model away from half filling,[12] except that our model also contains the hard-core repulsion between quasi-particles. We expect to find an antiferromagnetic instability for small δ, and that some form of superconducting pairing will be found in the Fermi-liquid state. Work in this direction is in progress. Finally, we emphasize that while this paper focused on solution in the $4t \ll D$

limit, the model can be solved numerically for arbitrary t/D. This will be necessary to compare with the experiment, since our interpretation of $\varepsilon_p - \varepsilon_d^o = 2\,\text{eV}$ and $\varepsilon_p - \varepsilon_d \approx 0.5\,\text{eV}$ puts us in the intermediate coupling $4t_{pd} \approx D$ regime.

ACKNOWLEDGEMENTS

This work is supported by the National Science Foundation under Grant No. DMR-85-21377 and through the Materials Research Laboratory under Grant No. DMR-84-18718

REFERENCES

1. C. M. Varma, S. Schmitt-Rink, and E. Abrahams, *Sol. State Comm.*, **61**, 681 (1987).
2. V. Emery, *Phys. Rev. Lett.*, **58**, 2794 (1987).
3. J. Hirsch, *Phys. Rev. Lett.*, **59**, 228 (1987).
4. Z. Shen, J. W. Allen, *et al.*, Xerox PARC preprint.
5. J. Zaanen, G. A. Sawatzky, and J. W. Allen, *Phys. Rev. Lett.*, **55**, 418 (1985).
6. A. Auerbach and K. Levin, *Phys. Rev. Lett.*, **57**, 877 (1986).
7. A. J. Millis and P. A. Lee, *Phys. Rev. B*, **35**, 3394 (1987).
8. J. Orenstein, *et al.*, AT&T Bell Labs preprint.
9. K. Kamaras, *et al.*, Univ. of Florida preprint.
10. C. M. Varma, S. Schmitt-Rink, and E. Abrahams, *Proceedings of the International Conference on Superconductivity*, ed., S. E. Wolf and V. Z. Kresin, Plenum, New York, 1987, and this volume.
11. M. Lavagna, A. J. Millis, and P. A. Lee, *Phys. Rev. Lett.*, **58**, 266 (1987).
12. B. Baskaran, Z. Zou, and P. W. Anderson, *Sol. State Comm.*, **63**, 973 (1987) and this volume.

J. E. Hirsch
Department of Physics
University of California, San Diego
La Jolla, California 92093

Pairing Mechanism in Oxide Superconductors: Insights From Small Systems Calculations

A useful way to learn about the pairing mechanism that is responsible for high T_c superconductivity is to study properties of model Hamiltonians on small systems. The goal is to find the simplest model that can describe the essential physics of high T_c superconductivity. We have used Monte Carlo simulation and exact diagonalization techniques to study properties of systems of up to 64 sites. Our results show that spin fluctuations and other spin related mechanisms induced by a Hubbard on-site repulsion U are not likely to give rise to pairing, neither in one nor in multiple band models. In contrast, charge fluctuations in a model with both strong U and V (repulsion between Cu and O) are shown to give rise to pairing and it is suggested that this model provides a plausible mechanism for high T_c superconductivity.

The mechanism giving rise to high temperature superconductivity is likely to be contained in a fairly simple tight binding Hamiltonian. A useful way to determine which the relevant Hamiltonian and mechanism are is to study properties of model Hamiltonians exactly on finite lattices. Although this may not be enough for quantitative answers, it should establish the essential qualitative features, particularly since the size of the Cooper pairs is believed to be only a few lattice spacings in these systems. Such a procedure will be free of uncontrolled approximations.

We have been engaged in such a systematic study,[1] starting from the single band Hubbard model in two dimensions and continuing with a three-band model to describe the CuO_2 planes in the oxide superconductors. We will summarize here our results from Monte Carlo simulations and exact diagonalization studies. Our results suggest that an on-site repulsion U is not likely to lead to superconductivity (at least in the interesting temperature range) neither in the single nor in the multiple band model. In contrast, the three-band model with U and a nearest-neighbor repulsion V is found to exhibit a pairing mechanism, and we argue that it is a plausible candidate as the "minimal" model to explain high T_c. Varma, et al.[2] have first suggested that the interaction V plays an essential role in high T_c superconductivity.

We start by considering the two-dimensional Hubbard model

$$H = \sum_{\langle i,j \rangle} t(d_{i\sigma}^\dagger d_{j\sigma} + \text{h.c.}) + U \sum_i n_{i\uparrow} n_{i\downarrow} \tag{1}$$

on a square lattice, representing an effective Hamiltonian for the holes on the Cu^+ $(3d^{10})$ ions. Anderson[3] has proposed that this model contains the essential physics of high T_c. In connection with heavy fermion superconductivity, and based on the strong coupling effective Hamiltonian derived from Eq. (1), we suggested some years ago that nearest-neighbor singlet pairs induced by the on-site repulsion U would condense into an anisotropic singlet superconducting state away from the half-filled band.[4] Weak coupling calculations supported this picture and predicted the anisotropic singlet state to be d-wave like close to the half-filled band.[5-6] Recent mean-field calculations[7] based on the "resonating valence bond" concept[3] also predicted nearest-neighbor singlet pairs to become Cooper pairs in a superconducting state away from 1/2-filling at low temperatures.

Unfortunately, recent Monte Carlo simulation results[8] suggest that the repulsive Hubbard model is not superconducting, at least in the interesting temperature range. We have computed pair susceptibilities

$$P_r = \int_0^\beta d\tau \langle \Delta_r(\tau) \Delta_r^\dagger(0) \rangle \tag{2}$$

as well as pair correlation functions

$$C_r = \langle \Delta_r \Delta_r^\dagger \rangle \tag{3}$$

with

$$\Delta_r = \sum_k f_r(k) C_{k\uparrow} C_{-k\downarrow} \tag{4}$$

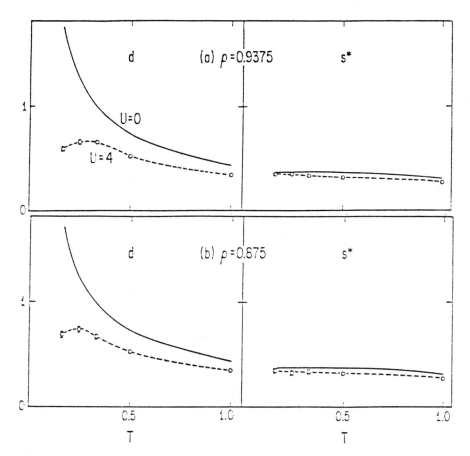

FIGURE 1 Pair susceptibilities for d-wave and extended s-wave for a 4 × 4 Hubbard model versus temperature for $U = 0$ (full line) and $U = 4$ (open circles connected by dashed line). Similar results are obtained for 6 × 6 and 8 × 8 lattices.

f_r is the form factor for the pair, and we have considered a d-wave-like state

$$f_d(\vec{k}) = \cos k_x - \cos k_y \tag{5}$$

and an extended s-wave-like state

$$f_s(\vec{k}) = \cos k_x + \cos k_y \tag{6}$$

among others. We have studied lattices of sizes 4 × 4, 6 × 6 and 8 × 8, temperatures from 1 down to 0.08 (in units of t), interaction strengths from $U = 0$ to 16 and band fillings $\rho = 1$ (half-filled) to $\rho = 0.5$. Our results show that even though pair *correlation functions* can be enhanced by U,[4] U always suppresses all pair *susceptibilities* compared to the non-interacting case. An example is shown in Fig. 1. In contrast, Fig. 2 shows the large enhancement of the $\vec{q} = (\pi, \pi)$ magnetic

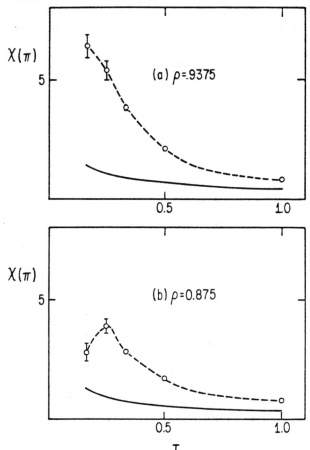

FIGURE 2 Magnetic susceptibility for wave-vector $\vec{q} = (\pi, \pi)$ for a 4×4 Hubbard model versus temperature for $U = 0$ (full line) and $U = 4$ (open circles connected by dashed line). $\rho = .9375$ and $\rho = .875$.

susceptibility produced by U close to the half-filled band. These results suggest that antiferromagnetic spin fluctuations suppress rather than enhance the tendency to superconductivity, at least in this temperature range. Furthermore, in an exact diagonalization study on 8-site lattices,[9] where we considered all possible pair wave functions that fit in such a lattice and studied a wide range of temperatures and interaction strengths again we found consistent suppression of all pairing susceptibilities with U.

Consider for comparison the attractive Hubbard model, which we would expect to go superconducting for any $U < 0$. Fig. 3 shows the s-wave pair susceptibility for $U = -2$, $\rho = 0.5$ on a 4×4 lattice. Note the large enhancement of the pair susceptibility compared to the non-interacting case, even at quite high temperatures. Mean field theory for this case predicts $T_c = 0.1t$, probably approximately correct, and a coherence length $\xi \sim 40$ lattice spacings. Note how our Monte Carlo results, even

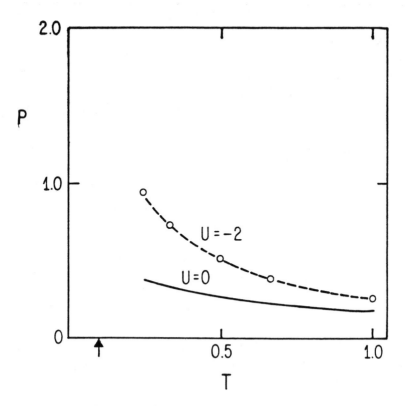

FIGURE 3 s-wave pair susceptibility for an attractive Hubbard model on a 4 × 4 lattice, $\rho = 0.5$, versus temperature for $U = 0$ (full line) and $U = -2$ (dashed line). The arrow denotes the mean field transition temperature.

on a lattice substantially smaller than ξ in linear dimension and at temperatures higher than T_c still show qualitatively the correct behavior.

The relevant energy scales in the repulsive Hubbard model are the bandwidth $W = 8t$, the Hubbard repulsion U and the antiferromagnetic exchange $J = 2t^2/U$. Approximate theories for superconductivity predict T_c's of the same order of magnitude as J,[7] and a maximum $T_c \sim 0.1t$.[10] Our results show that there is no superconductivity at those temperatures. If we assume an effective bandwidth of 2 to 3 eV in the oxide superconductors, our Monte Carlo studies covered temperatures down to approximately 200 K. Given that we have explored a wide range of parameters, that we would not expect nature to have found the absolute optimal set of parameters in $YBaCu_3O_{7-\delta}$, and that we expect to see enhancement of pair susceptibilities even well above T_c, we conclude that the Hubbard model does not contain the essential physics to describe high T_c superconductors.[11]

The next step is to study a two-dimensional lattice appropriate to describe the CuO_2 planes. We assume a Hubbard repulsion U on the Cu sites and an oxygen level position ϵ with respect to the Cu level. Emery has suggested that a spin-exchange pairing mechanism should work in such a lattice but not in the single

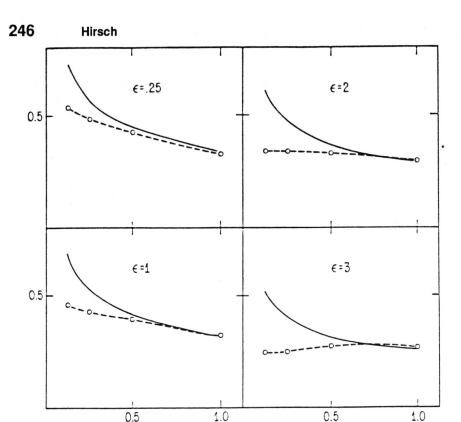

FIGURE 4 Nearest-neighbor Cu-O pair susceptibility versus temperature for $U = 0$ (full line) and $U = 4$ (dashed line) for various values of ϵ. $\rho \sim 1$.

band model.[12] We have suggested a pairing mechanism for two oxygen holes on this lattice due to the "strings" of overturned Cu spins created when a O hole travels through the lattice.[13] Such a string would generate an attractive potential between oxygen holes that increases linearly with their separation for time scales short enough that spontaneous exchange of Cu spins can be neglected.[14]

While these mechanisms appear plausible and may be realizable in some extreme parameter regime, our simulations show that they do not seem to work in the intermediate coupling regime which we believe is appropriate to the oxide superconductors. We have performed Monte Carlo simulations on 48-site lattices, at and close to the half-filled band case, for several values of ϵ and U, and calculated pair susceptibilities for O-O, O-Cu and Cu-Cu pairing. The Cu-Cu pair susceptibilities behave similarly as for the single band case, examples for Cu-O and O-O pairing are shown in Figs. 4 and 5 for $\rho \sim 1$. Results are similar for a doped case ($\rho \sim 1.25$). Despite the fact that the O-O pair susceptibility is enhanced by U, the fact that the enhancement is temperature independent suggests it is a trivial effect. In fact, it comes from a renormalization of the Cu level position which yields an effective $\epsilon' = \epsilon - U\langle n_{cu}\rangle/2$. In taking this into account we find again that the net effect of U is to suppress the pair susceptibilities. Once again, spin susceptibilities are greatly

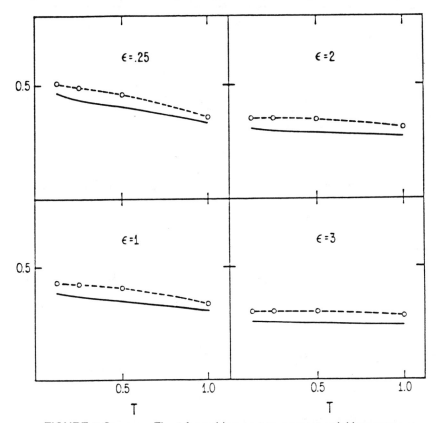

FIGURE 5 Same as Fig. 4 for pairing on two nearest-neighbor oxygens.

enhanced by U, as shown in Fig. 6. These are rapidly suppressed as ρ is moved away from 1.

These results suggest that a model with just an on-site Hubbard repulsion U is not adequate to explain high T_c. It is natural then to move on to a more realistic model that includes also the nearest-neighbor Cu-O Coulomb repulsion V. Varma, et al.[2] have emphasized that this is an important interaction. Indeed, simple strong coupling arguments suggest to us that this is the simplest model that will exhibit superconductivity induced by purely repulsive interactions. The Hamiltonian is:

$$H = \sum_{(i,\ell)} t(d_{i\sigma}^\dagger C_{\ell\sigma} + \text{h.c.}) + (\epsilon - \mu)\sum_{\ell\sigma} C_{\ell\sigma}^\dagger C_{\ell\sigma} - \mu\sum_{i\sigma} d_{i\sigma}^\dagger d_{i\sigma} \qquad (7)$$

$$+ U\sum_i n_{i\uparrow}n_{i\downarrow} + U_p\sum_\ell n_{\ell\uparrow}n_{\ell\downarrow} + V\sum_{(i,\ell)} n_i n_\ell + V_{OO}\sum_{(\ell\ell')} n_\ell n_{\ell'} \qquad (8)$$

where $d_{i\sigma}^\dagger$ creates a Cu hole at site i, and $C_{\ell\sigma}^\dagger$ an O hole at site ℓ. U_p is an on-site repulsion for the oxygen holes, which we need to include for consistency. We have also included a nearest-neighbor O-O repulsion V_{OO} for reasons discussed below. It is reasonable to assume $U > U_p \gtrsim V > V_{OO}$.

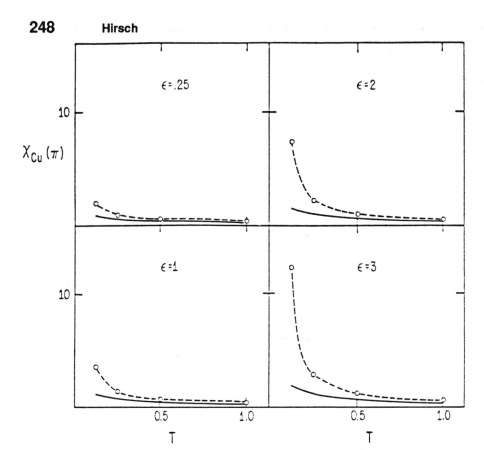

$\chi_{Cu}(\pi)$

FIGURE 6 Cu-Cu spin susceptibility for wave vector $\xi = (\pi, \pi)$ for $U = 0$ (full line) and $U = 4$ (dashed line) for various values of ϵ. $\rho \sim 1$.

Consider the extreme situation where $U \to \infty$, $\epsilon \gg t$ and there is one hole per Cu atom. Adding one extra hole (on an oxygen site) will cost energy $E_1 = \epsilon + 2V$. Now consider the energy of a state with two added oxygen holes. If they are far away, the energy of the state is:

$$E_2' = 2E_1 = 2\epsilon + 4V$$

If put on the same site as the first O hole, however, the neighboring Cu holes will be pushed away, as shown in Fig. 7, giving a state of energy:

$$E_2 = 2V + 4\epsilon + U_p \tag{10}$$

Thus, this situation is favorable if the "effective U"

$$U_{\text{eff}} = E_2 - 2E_1 = -2V + 2\epsilon + U_p \tag{11}$$

is negative. The possibility of superconductivity arising from such "negative U centers" has been discussed elsewhere.[15] In a strong coupling expansion in t, it is

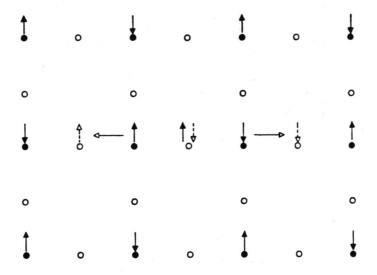

FIGURE 7 Addition of 1 and 2 oxygen holes to the state with one hole per Cu atom. Cu atoms are denoted by black circles, O atoms by open circles. On adding the second hole to the oxygen on the center of the picture the neighboring Cu holes are pushed away onto neighboring oxygens (dashed lines). The difference in energy between putting the two O Holes on the same versus on distant oxygens is $U_{\text{eff}} = -2V + 2\epsilon + U_p$.

easy to convince oneself that superconductivity will occur from Bose condensation of these tightly bound pairs. Finally, a small O-O repulsion $V_{\text{OO}} \sim V/\sqrt{2}$ is realistic and necessary to avoid real condensation (phase separation) of the O holes.

The smallest system that will exhibit this phenomenon is a 6-site Cu-O chain (3Cu, 3O alternating). Here we can examine this effect away from the strong coupling limit. Calculations of U_{eff} show that it is negative over a substantial parameter range; examples are shown in Figs. 8 and 9. Results of pair susceptibilities also show enhancement as V is increased. These results suggest that *both* U and V need to be appreciable, and that the mechanism only works for a limited range of ϵ. Calculations on larger clusters are in progress[16] and should provide detailed information on the feasibility of this mechanism.

In summary, our numerical results have shown that models with only on-site repulsion U are unlikely to exhibit superconductivity. In contrast, including a nearest-neighbor repulsion V appears to provide a plausible pairing mechanism. The Cu-O lattice structure here is essential, as it is easy to convince oneself that at least in strong coupling the mechanism would not work in a single band model. Our results showed that a large U on the Cu ions is necessary for this mechanism, which is encouraging as the existence of antiferromagnetism in these systems suggest that U is appreciable. This mechanism predicts s-wave superconductivity, in agreement with recent experiments,[17] and appears to contain several features that are specific

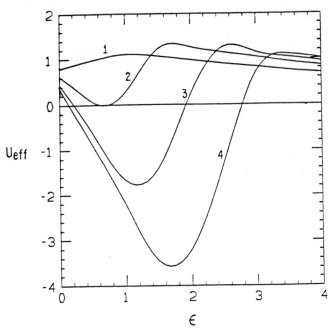

FIGURE 8 U_{eff} for a 6-site chain as defined in the text versus ϵ for 1: $U = 4$, $V = U_p = 2$; 2: $U = 8$, $V = U_p = 4$; 3: $U = 12$, $V = U_p = 6$; 4: $U = 16$, $V = U_p = 8$.

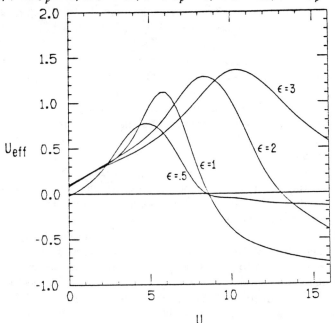

FIGURE 9 Same as Fig. 8 for $U_p = V = 4$ and $\epsilon = .5, 1, 2, 3$.

to the high T_c oxides and absent in other materials. Thus, a 2-dimensional Cu-O lattice with U, V, U_p, and V_{OO} interactions may be the "minimal model" that contains the essential physics of high T_c.

ACKNOWLEDGEMENTS

I am grateful to H. Q. Lin, S. Tang, G. Bickers, E. Loh and D. Scalapino for discussions and collaboration. Stimulating conversations with V. Emery, D. Haldane and C. Varma are also acknowledged. This work was supported by the National Science Foundation under Grant No. DMR-85-17756. Computations were performed at the Cray XMP of the San Diego Supercomputer Center. I am grateful to AT&T Bell Laboratories and Cray Corporation for financial support.

REFERENCES

1. Different portions of the work discussed here were performed in collaboration with H. Q. Lin, D. J. Scalapino, and S. Tang.
2. C. M. Varma, S. Schmitt-Rink, and Elihu Abrahams, *Solid State Commun.*, **62**, 681 (1987).
3. P. W. Anderson, *Science*, **235**, 1196 (1987).
4. J. E. Hirsch, *Phys. Rev. Lett.*, **54**, 1317 (1985).
5. D. J. Scalapino, E. Loh, and J. E. Hirsch, *Phys. Rev. B*, **34**, 8190 (1986).
6. J. Miyake, S. Schmitt-Rink, and C. Varma, *Phys. Rev. B*, **34**, 6554 (1986).
7. G. Baskaran, Z. Zou, and P. W. Anderson, *Sol. St. Comm.*, **63**, 973 (1987); A. E. Ruckenstein, P. J. Hirschfeld, and J. Appel, *Phys. Rev. B*, **36**, 857 (1987) and this volume.
8. J. E. Hirsch and H. Q. Lin, to be published.
9. H. Q. Lin, J. E. Hirsch, and D. J. Scalapino, to be published.
10. H. Fukuyama and K. Yosida, *Jpn. J. Appl. Phys.*, **26**, L371 (1987).
11. While our results demonstrate that the Hubbard model does not exhibit superconductivity at temperatures comparable to J, we cannot exclude the possibility of superconductivity at exponentially smaller temperatures. Recent $T = 0$ variational calculations on the Hubbard model [C. Gros, R. Joynt, and T. M. Rice, *Phys. Rev B*, **36**, 381 (1987); C. Gros, preprint, suggest d-wave airing in the ground state.
12. V. J. Emery, *Phys. Rev. Lett.*, **58**, 2794 (1987) and this volume.
13. J. E. Hirsch, *Phys. Rev. Lett.*, **59**, 228 (1987).
14. B. Shraiman and E. Siggia, preprint, have recently studied this problem in a single-band model.

15. J. E. Hirsch and D. J. Scalapino, *Phys. Rev. B*, **32**, 5639 (1985).
16. J. E. Hirsch, E. Loh, D. J. Scalapino, and S. Tang, unpublished.
17. D. R. Harshman, *et al.*, *Phys. Rev. B*, **36**, 2386 (1987).

Subject Index

ac Josephson effect, 20
acoustic plasmon branch, 93
acoustic plasmons, 93
AFM *vs.* RVB, 168
Ag, 22
Al, 22
anisotropic conductivity, 175
anisotropy, 22
    of excitation spectrum
    effects on infrared and tunneling data, 40
antiferromagnetic order
    in Hubbard model, 143
antiferromagnetism
    and oxygen vacancies, 107
    in La_2CuO_4, 108
    of stoichiometric oxides, 184
antiferromagnets
    conjectured renormalization group flows, 205
    possible experimental observations of "strong coupling phase", 205
    triangular lattice, 205
attractive Hubbard model, 244
auxiliary Bose fields, 139

background conductance
    in tunneling, 174
band structure calculations
    and antiferromagnetism in oxides, 184
$BaPb_{1-x}Bi_xO_3$, 15,100
barium titanate, 17
Bethe Ansatz solution for the Heisenberg antiferromagnet, 186
bipolaron formation, 104
bipolaron models, 212
bond asymmetry
    effects on phonon mechanism, 80
    static model, 86
    term, 81
Bose condensation of bipolarons, 137
    in RVB state, 131
    of bipolarons, 104
    of slave bosons, 139
boson hole solitons
    scattering by spinons, 176
Ce, 21
chains and planes
    coupling between, 30
chains, 17, 19
charge transfer excitations (CTE), 54
charge transfer resonance resonance mechanism, 211
charge transfer resonances
    longitudinal, 220
    transverse, 221
    comparison of experiment with transverse mechanism, 222
    $Cu^{3+}O^{--} \rightarrow Cu^{+2}O^-$, 217
    width, 219
cluster calculations
    in oxides, 185
Co, 21
coherence length, 20, 22
    band structure estimates, 49
    from specific heat, 30
    in slave boson mean field theory, 142
Coleman's theorem, 200
compressibility, 239
conductivity
    in plane, 176
Coulomb repulsion
    role in determining T_c, 92
    in polaron model, 104
covalent-metallic picture, 138
critical current density, 20
critical currents, 14

anisotropy of, 22
critical fields, 14
    anisotropy of, 22
    upper, 49
crossover between BCS-like and Bose condensation, 142
crystal structure, 17
Cu $d_{x^2-y^2}$-O(1) p$_\parallel$, 44
Cu^{3+}, 20, 217
    absence in La$_{2-x}$Ba$_x$CuO$_4$, 185
    absence in YBa$_2$Cu$_3$O$_7$, 185
Cu
    role in polaron model, 105
d-wave pairing
    in Hubbard model, 144
d-wave superconductivity by exchange of anti-paramagnons, 190
    in bond asymmetry models, 88
d-wave-like state, 243
dangling bonds
    in RVB state, 158
de Almeida-Thouless line, 22
Debye-Waller factors, 100
density of states, 226
diamagnetism, 22
dielectric constant
    in Y-Ba-Cu-O system, 63
dielectric screening
    analysis in determining T_c, 92
domains
    in bond asymmetry model, 84
elastic constants, 100
electrical conductivity
    frequency dependent, 100
electromagnetic properties, 181
electron tunneling, 24
electron-electron attraction
    reduction by large boson frequency, 92
electron-phonon interaction, 11
    stabilizing RVB state, 155
    estimates for YBa$_2$Cu$_3$O$_{7-\delta}$, 58
electron-phonon mechanism, 79, 137
electron-phonon models, 212
elemental substitution, 21
elementary electronic excitations
    in RVB state, 156
Eliashberg equation, 96
energy bands
    for ordered vacancy model of LaCu$_2$O$_{4-x}$, 109
energy loss function

for Y-Ba-Cu-O system, 66
equation of motion method, 121
exact diagonalization, 242
excitation spectrum
    in slave boson theory of Hubbard model, 142
exciton binding energy
    estimates for, 72
exciton broadening, 73
excitonic condensate, 71
    conditions for, 72
excitonic enhancement of superconductivity, 69
    experimental consequences, 72
excitonic mechanism
    and oxygen vacancies, 107
excitonic nature
    of excitations in oxides, 185
excitons
    in $YBa_2Cu_3O_{7-\delta}$, 58
extended Hubbard model, 228
    limit of infinite U on copper sites and zero U on oxygen sites, 235
fermion-boson field theory
    of RVB state, 163
        insulating state, 166
            spin soliton, 166
        meaning of the operators, 165
        relation to true electron operators, 165
ferroelectric Curie temperature, 17
FLAPW
    energy band calculation, 44
flourscence
    at T_c for excitonic mechanism, 76
Flux phase, 150
flux quantization, 20, 169
    in RVB state, 156
gap for spin excitations, 148
gapless Fermi surface, 148
gauge symmetry
    in slave boson formulation double occupancy constraint, 141
glass
    superconducting, 22
Gutzwiller approximation, 164
Gutzwiller projection, 131, 134
Gutzwiller state
    by variational Monte Carlo methods, 183
Gutzwiller wavefunction, 186
    absence of superconductivity at half filling, 187
    below half filling
        by Monte Carlo methods, 187
            superconductivity, 188

physical origin of superconductivity, 188
Hall Effect, 20
Hartree-Fock factorization
in slave boson mean field theory, 139
heat capacity, 20
heavy fermion systems , 164, 213, 227, 190
Heisenberg antiferromagnet in two dimensions, 187
Heisenberg Hamiltonian, 194
Bethe ansatz solution, 197
including spin-lattice interaction, 158
Heisenberg model
dimerized ground state in large n limit, 203
large n limit
effective action, 149
gauge invariance, 149
ground state, 149
path integral method, 149
valence bonds in, 148
s=1/2, 148
Heisenberg-Hubbard model
large n limit, 151
charge density waves, 151
fluctuations in phase of order parameter, 151
impossibility of superconductivity, 151
phase diagram, 153
Heisenberg-Peierls model, 194
LSM theorem, 199
Hg, 8
holon-spinon elastic scattering, 177
Holstein-Primakoff approximation, 201
honeycomb lattice antiferromagnet, 194
unique ground state and a gap, 195
Hopf topological term, 203
hopping matrix element t
effect on T_c, 40
Hubbard model, 242
Hubbard model
effective interaction parameters, 138
extended
elimination of Cu(3d) states, 229, 231
effective attractive coupling, 230
gap equation, 231
elimination of the oxygen sites, 229
Fermi liquid picture, 237
holes on O sites, 229
interaction due to slave boson exchange, 239
intermediate coupling, 230
large n expansion, 236

        leading term in $1/N$, 239

        mean field theory, 236

        Monte Carlo calculations, 241

        optical conductivity, 237

        two band case, 245

        with Cu-O repulsion V

          argument for pairing, 248

    large n limit, 148

    mean field theory using slave bosons, 137

    Monte Carlo calculations at finite temperature, 241

    relevance to oxide superconductors

      evidence from band structure, 214

      evidence from cluster calculations, 215

    single-band at large U, 184

      relevance if holes are on O-sites, 185

      relevance to superconductivity, 184

Hubbard-Peierls models, 157

Hubbard-Stratonovich transformation, 149

infrared absorption

    as test of excitonic enhancement, 72

infrared quenching

    as evidence for excitonic effects, 74

infrared reflectance

    analysis for excitonic effects, 73

infrared spectra, 71

interlayer coupling

    role of, 35

interlayer tunneling process, 177

isotope effect, 21, 100

    in bond asymmetry model, 83

    in $La_{2-x}Ba_xCuO_4$, 30

    in polaron model, 104

    in $YBa_2Cu_3O_{7-x}$, 30

itinerant electron models

    LSM theorem, 199

Josephson coupling

    stabilising phase fluctuations, 36

K_2NiF_4, 17

Kite phase, 153

Kosterlitz-Thouless process, 181

La sites

    electronic structure, 45

La_2CuO_4, 147

    effects of doping, 229

    monoclinic, 108

    orthorhombic distortion, 153

    orthorhombic, 108

    p-bands, 185

undoped, 185

$La_{2-x}Ba_xCuO_4$
    absence of Cu^{3+}, 185
    holes on oxygen sites, 185

$La_{2-x}M_xCuO_4$, 17
    band structure, 54
    charge fluctuations between in plane Cu atoms, 54
    frozen phonon calculations, 54

$La_{2-\delta}Sr_\delta CuO_4$, 148

Lagrange multiplier
    in slave boson mean field theory, 139

Landau damping, 219
    of plasmons, 95

lattice parameter
    variations due to Lifshitz transition, 47

LCAO method
    applied to $YBa_2Cu_3O_7$, 64

Lieb-Schultz-Mattis theorem , 148, 151, 193, 195, 197
    alternating interactions, 197
    higher dimensional generalizations
        honeycomb lattice, 199
        three dimensions, 199
        triangular lattice, 199
        two dimensions, 197
    itinerant electron models, 199
    more general Hamiltonians, 197
    nature of strong-coupling phase in non-linear σ model, 203
    relation to RVB state, 203

Lifshitz transition, 43

linear resistivity, 102

linear temperature dependent resistivity, 168

$LiTi_2O_4$, 15

local density calculations
    band structure, 43

localization length, 230

logarithmic singularity
    in 2D band structure, 81

Lu, 21

magnetic models, 213

magnetic susceptibility, 100
    in slave boson theory of Hubbard model, 142

Mattheiss's effective tight-binding model, 229

$MBa_2Cu_3O_{7-x}$, 15

Meissner-Ochsenfeld Effect, 16

Mermin-Wagner theorem, 200

metallic hydrogen, 92

Metallic oxides, 10

Migdal theorem, 92

Monte Carlo calculations
        finite temperature on small systems, 241
Monte Carlo method
        variational for Hubbard model, 183, 186
Mott insulator, 186
Nb_3Ge, 8
nearest-neighbor repulsion V, 242
neutron scattering, 190
Ni, 21
normal tunneling, 178
nuclear resonance on ^{63}Cu, 30
O atoms
        onsite interaction, 185
O^-, 217
O_2^{2-}, 217
$O(2p)$ states
        holes as charge carriers, 227
optical absorption, 235
optical conductivity, 101, 237
optical studies, 25
organic superconductors, 227
orthorhombic distortion
        in La_2CuO_4, 153
orthorhombic tetragonal structural transition, 148, 18
        in $YBa_2Cu_3O_{7-x}$, 56
oxygen stoichiometry, 17
oxygen vacancies, 17
        and antiferromagnetism, 107
        and excitonic mechanism, 107
        band structure with, 107
        charge density near, 110
        enhancement of oxygen hole exchange, 125
        in $LaCu_2O_4$, 107
        in $YBa_2Cu_3O_{7-x}$, 108
        model for pairing mechanism, 118
        model for pairing mechanism
            solution, 121
        narrow band formed by, 110
        ordering of, 18
        pairing mechanism, 115
            percolative picture, 117
        screening, 116
p-wave superconductivity
        in bond asymmetry models, 88
pair correlation functions, 242
pair susceptibilities, 242
Peierls phase, 150
perovskite compounds, 13

phase fluctuations
    nature of, 36
phonon exchange model
    including bond asymmetry, 84
        isotope effect, 86
phonon exchange, 227
phonons
    role of, 29
photoemission spectroscopy, 98
plaquettes, 160
plasma frequency, 230
    for Y-Ba-Cu-O system, 67
plasmon mechanism of superconductivity, 91
    effects of alloying, 93
    in two dimensional systems, 93
    in two band model
        damping, 96
        pairing parameter, 96
plasmon spectrum
    in two band model, 94
polaron band, 100
polaron formation, 102
polaron Hamiltonian, 104
polaron model, 99
polaron superconductivity, 99
polyacetylene
    solitons compared to dangling bonds in RVB, 159
positive temperature coefficient of electrical resistivity, 17
positron annihilation, 116
    temperature dependence from vacancy model, 122
Pr, 15, 21
pseudo Fermi surface, 142
pseudofunction method
    for O vacancies in $LaCu_2O_4$, 109
quasiparticle picture, 138
quasiparticle statistics
    in RVB state, 159
residual 3D interactions, 47
resistivity perpendicular to the Cu-O plane, 178
resistivity
    anisotropy of, 22
Resonant photoemission, 185
resonant-valence-bond model, 228
resonating valence bond state
    excitations, 155
    mean field theory, 129
        insulating phase, 132
        superconducting phase, 133
    near half filling, 134

Rietveld refinement techniques, 18
rigid band approximation, 47
RVB state
     by variational Monte Carlo methods, 183
     charge e bosons, 163
     neutral fermion excitation spectrum, 163
     relation to superconductivity, 159
     tunneling and anisotropic resistivity, 173
s-wave-like state, 243
saddle point singularity(SPS), 43
sigma-model
     mapping from antiferromagnet, 194, 200
       Lagrangian density, 200
       perturbative spectrum, 201
       relation to RVB state, 203
       square lattice, 201
         additional fields, 202
singlet pairing, 30
slave boson method, 164
     in extended Hubbard model, 236
     relation to soliton holes, 164
soft phonon mechanism, 93
soliton-antisoliton pairs, 156
solitons in RVB state
     charged, 159
     effective mass, 159
Sommerfeld χ/γ, 100
sound velocity, 100
specific heat, 100
     electronic contribution
       interaction effects, 48
     in narrow band polaron model, 102
     in RVB model, 132
     in slave boson theory of Hubbard model, 142
     linear low-temperature, 148
     spinons, 167
       effects of magnetic impurities, 168
spin configurations
     local
       strong coupling to as mechanism, 227
spin Peierls phase, 195
spin-glass-like behavior, 22
spin-Peierls phase
     relation to RVB, 157
spin-wave coupling constant, 194
spinons, 163
square root singularity
     effects of next nearest neighbors, 82

from bond asymmetry, 81
including Cu atoms, 82
Stoner factors
in $YBa_2Cu_3O_7 - x$, 57
strontium and lead titanates, 17
SU(n) generalizations of Heisenberg model
LSM theorem, 199
superexchange in transition-metal salts, 138
superexchange, 139
between copper spins, 236
superlattice splitting
in O vacancy model, 110
susceptibility
in narrow band polaron model, 102
in RVB state, 133
Tb, 15, 21
thermistor, 17
thermopower, 102, 178
three-band model, 242
tight binding model
corrections to mean field theory, 214
phase diagram, 215
titanate ceramics, 17
topological defects
in RVB state, 158
topological excitations, 155
topological long-range order, 155
in RVB state, 158
topological solitons, 156
transition metal oxides
tight binding model, 213
corrections to one electron theory, 214
tungsten bronzes, 97
tunneling and anisotropic resistivity, 173
tunneling between layers, 177
as origin of superconductivity, 181
tunneling centers, 102
tunneling, 20
twin boundaries
relation to glass-like behavior, 23
twitch transition, 163
two band model, 213
two dimensional nature of the electronic structure, 47
two fluid model for chains and planes, 31
two-level systems, 20
ultrasonic attenuation, 100
Umklapp processes, 176
Uniform phase, 153

valence bond creation operator, 149
van Hove saddle point singularity, 44
  effects on T_c, 48
virtual crystal approximation, 47
x-ray-absorption near-edge study of $La_{2-x}Ba_xCuO_4$, 229
Yb, 21
$YBa_2Cu_3O_7$
  absence of Cu^{3+}, 185
  carrier density, 20
  holes on oxygen sites, 185
$YBa_2Cu_3O_{7-x}$
  parameters, 20
  infrared spectra, 71
  band structure, 55
  chains vs plains, 54
  charge density, 55
  density of states, 55
  optical properties, 63
  tetragonal phase, 54
Young's modulus, 100
Zou-Anderson-Wheatley theory, 176